W9-DED-005

Employment Dynamics
in Rural Europe

Employment Dynamics in Rural Europe

Edited by

Ida J. Terluin and Jaap H. Post

Agricultural Economics Research Institute (LEI)
The Hague
The Netherlands

CABI *Publishing*

CABI *Publishing* **is a division of CAB** *International*

CABI Publishing
CAB International
Wallingford
Oxon OX10 8DE
UK

CABI Publishing
10 E 40th Street
Suite 3203
New York, NY 10016
USA

Tel: +44 (0)1491 832111
Fax: +44 (0)1491 833508
Email: cabi@cabi.org
Web site: http//www.cabi.org

Tel: +1 (212) 481 7018
Fax: +1 212 686 7993
Email: cabi-nao@cabi.org

A catalogue record for this book is available from the British Library, London, UK.

Library of Congress Cataloging-in-Publication Data
Employment dynamics in rural Europe / edited by Ida J. Terluin and Jaap H. Post.
    p.cm.
  Includes bibliographical references and index.
  ISBN 0-85199-499-7 (alk. paper)
  1. Manpower policy, Rural--European Economic Community countries--Case studies.
2. European Economic Community countries--Economic conditions. I. Terluin, Ida J. II. Post, Jaap H.

HD5710.85.E86 E47 2000
331.1'091734--dc21

00-044466

ISBN 0 85199 499 7

Printed and bound in the UK at Cromwell Press, Trowbridge
from copy supplied by the editors

Contents

Contributors

Olli Aulaskari is Research Manager at the Finnish Regional Research FAR, Sonkajärvi, Finland.

Jean-Pierre Bertrand is Senior Researcher at the Department of Rural Economics and Sociology (DESR) in the Research Unit on Environment, Techniques, Societies and Economic Policies (STEPE) of the Institut National de la Recherche Agronomique (INRA), Nogent-sur-Marne, France.

Pierre Dupraz is Researcher at the Department of Economics of the Institut National de la Recherche Agronomique (INRA), Rennes, France.

Angelos Efstratoglou is Research Associate at the Department of Economics and Regional Development of the Panteion University of Athens, Greece.

Sophia Efstratoglou is Professor at the Department of Agricultural Economics and Rural Development of the Agricultural University of Athens, Greece.

Roberto Esposti is Researcher at the Department of Economics, Università degli Studi di Ancona, Italy.

Frans E. Godeschalk is Senior Research Assistant in the General Economics and Statistics Division, Agricultural Economics Research Institute LEI, The Hague, The Netherlands.

Bruno Henry de Frahan is Professor of Agricultural Economics and Policy at the Agricultural Economics Research Unit of the Université Catholique de Louvain, Belgium.

Gemma Francés is Researcher at the Department of Applied Economics of the Universitat Autonoma de Barcelona, Spain.

Tuomas Kuhmonen is Managing Director of Fin-Auguuri Oy (Ltd), Vesanto, Finland.

Heino von Meyer is Principal Administrator at the OECD Environment Directorate, Paris, France. He was previously Head of PRO RURAL Europe, Wentorf (Hamburg), Germany.

Jaap H. Post is Head of the General Economics and Statistics Division of the Agricultural Economics Research Institute LEI, The Hague, The Netherlands.

Bernard Roux is Senior Researcher at the Department of Rural Economics and Sociology (DESR) of the Institut National de la Recherche Agronomique (INRA), Paris, France.

Jordi Rosell is Professor at the Department of Applied Economics of the Universitat Autonoma de Barcelona, Spain.

Danielle Schrobiltgen is Research Assistant for PRO RURAL Europe, Wentorf (Hamburg), Germany.

Franco Sotte is Professor of Agricultural Economics and Regional Economics at the Department of Economics, Università degli Studi di Ancona, Italy.

Ida J. Terluin is Senior Researcher in the General Economics and Statistics Division, Agricultural Economics Research Institute LEI, The Hague, The Netherlands.

Béatrice Van Haeperen is Lecturer of Labour Economics at the Department of Economics of the Université Catholique de Louvain, Belgium.

Lourdes Viladomiu is Professor at the Department of Applied Economics of the Universitat Autonoma de Barcelona, Spain.

Franz Weiß is Research Associate at the Department of Economics, Politics and Law at the University of Agricultural Sciences, Vienna, Austria.

Anita Wisselink is Official at the Directorate for Spatial Economic Policy of the Ministry of Economic Affairs, The Hague, The Netherlands.

Preface

This book is the result of a fruitful cooperation of over 20 researchers from nine EU member states in the RUREMPLO project. RUREMPLO is the acronym of 'Agriculture and employment in the rural regions of the EU'. The RUREMPLO project was supported under the EU research programme FAIR (CT 96 1766) and coordinated by the editors of this book. In this project, employment conditions and trends in rural regions of the EU have been analysed against the background of a downward trend in the agricultural labour force. In order to reveal and better understand the forces behind rural employment dynamics, an analysis was undertaken in two steps. First, basic socio-economic characteristics of all EU regions were analysed, then, in a second step, 18 in-depth case studies covering leading and lagging rural regions in nine EU member states were carried out. This resulted in a number of key messages towards enhancing the employment situation in rural regions.

In this book the issue of rural employment is examined from various viewpoints: first, a theoretical background of employment development in rural regions is discussed; second, a statistical analysis of employment development in all rural regions in the EU is given; third, employment dynamics in pairs of leading and lagging rural regions are analysed; and finally, some lessons for encouraging rural employment are given. The book is composed in such a way that together the chapters form an analysis of employment dynamics in rural Europe; however, each chapter acts as a building block of the whole and can also be read separately from the others.

We greatly acknowledge the assistance of Mei Li Tai and Urmila Koelfat in the layout of the manuscript.

Ida Terluin and Jaap Post

Employment Dynamics in Leading and Lagging Rural Regions

1

Ida J. Terluin and Jaap H. Post

Introduction

The creation of jobs and the fight against unemployment are political priorities in the EU. Traditionally, the agricultural sector has been an important source of employment in rural regions. However, agricultural employment has declined rapidly during the last few decades and there is evidence that the number of jobs in agriculture in the EU is about to halve in the next 20 years. The reduction of agricultural employment has major consequences for the employment situation of rural regions – especially when the share of agriculture in total employment is high – unless sufficient alternative jobs can be provided. This implies that industries and services are becoming increasingly important as sources of employment in rural regions.

To gain an insight into the means by which employment in industries and/or services in rural regions can be encouraged, it is useful to analyse employment performance in the recent past. Over the last decade, a number of rural regions in the EU experienced considerable employment growth in their non-agricultural sectors, while in other rural regions employment growth stagnated. We labelled the first group as 'leading regions' and the second group as 'lagging regions'. Once classified, the following research questions were asked: why do the two groups experience different employment dynamics, and, what lessons could lagging regions learn from leading ones? These questions were addressed in two steps: first, basic socio-economic characteristics of all EU regions were analysed, and in a second step, 18 in-depth case studies covering pairs of leading and lagging rural regions in nine EU member states were carried out.

Lessons towards stimulating employment creation in rural regions are helpful for all policy makers who are involved in employment affairs: at local, regional, national and EU level. At all these levels, measures are implemented towards encouraging employment and often policy makers of different levels are together involved in the implementation of policies, for instance in the EU structural policies. The more these policies are effectively targeted and efficiently implemented, the more the people living in rural regions can enjoy being employed. This will strengthen the viability of rural regions and contribute to the socio-economic cohesion among regions in the EU.

Plan of this book

The plan of this book is as follows. In the remainder of this chapter we give a broad outline of employment performance and of forces which have hampered and encouraged employment growth in rural regions in the EU since the beginning of the 1980s and we formulate some key messages for encouraging employment opportunities in rural regions. This can be seen as a preview of the book. In Chapter 2 the theoretical and methodological framework of employment dynamics is dealt with. It starts with an examination of the meaning of rurality, followed by a discussion of theories on economic development in rural regions. It is argued that the mixed exogenous/endogenous approach is useful for analysing employment dynamics in rural regions. Hence, this approach acted as a starting point for the design of a conceptual model: the field of force of a rural region. In Chapter 3 attention is paid to a statistical analysis of employment performance in all EU regions during the 1980s and first half of the 1990s. Then, in Chapters 4-12 the results of the case studies are discussed. In each of these chapters, an analysis is made of the most striking factors hampering and encouraging employment growth in each national pair of leading and lagging regions during the years 1980-1997. In the last chapter we put all nine leading case study regions into one group and the lagging case studies into another group, and report on the main similarities and differences in employment dynamics between both groups. As a final step, we derive a number of lessons, which leading and lagging rural regions can learn from each other with regard to employment creation.

Employment growth in rural regions of the EU, 1980-1995

In this book, the rural is viewed upon from the territorial approach, which implies that rural areas are expressed in terms of territorial entities, which cover a local or regional economy, include one or more small or medium sized cities and have a low population density (see Chapter 2). Such territorial entities were labelled as 'rural regions', and their sizes usually reflect that of labour market areas.

In order to examine employment growth during the 1980s and early 1990s, we used a division of 465 regions in the EU15. Based on the criterion of population density, we distinguished three types of rural regions: most rural regions, intermediate regions and most urban regions (see Chapter 3). Over 40% of the EU regions were classified as most rural, more than one-third as intermediate and nearly one-quarter as most urban.

Within the groups of most rural and intermediate regions, we made a further distinction into leading, average and lagging regions based on the performance of non-agricultural employment growth in the 1980s and early 1990s. A region is considered to be leading if the growth rate of non-agricultural employment was 0.5% points above the national growth rate; on the other hand, a region is

Table 1.1 Employment and population growth by types of regions, '1980-93' (% p.a)

Regions	Employment growth					Population growth
	Total	Agri-culture	Indus-try	Servi-ces	Non-agri-culture	
Leading most rural	0.8	-3.7	0.2	2.1	1.4	0.51
Leading intermediate	1.0	-3.6	0.3	2.4	1.6	1.09
Lagging most rural	-0.7	-4.6	-1.8	1.3	0.1	-0.06
Lagging intermediate	-0.7	-4.8	-2.0	1.0	-0.2	0.06
Most rural regions	0.0	-4.1	-0.8	1.7	0.8	0.26
Intermediate regions	0.4	-3.9	-0.6	1.7	0.8	0.52
Most urban regions	0.5	-3.3	-1.1	1.5	0.6	0.32
All regions	0.4	-3.9	-0.9	1.6	0.7	0.37

considered to be lagging if the growth rate of non-agricultural employment was 0.25% points below the national growth rate. By doing so, about one-third of the most rural and intermediate regions were classified as leading and a quarter as lagging. It has to be emphasized that the labels leading and lagging were only derived from employment performance, and that leading regions may be less successful with regard to other indicators. Moreover, it appears that the growth rate of employment can change considerably when using another period. This implies that if a region is labelled as lagging, this is not necessary a permanent situation but that it can change.

A statistical analysis of some socio-economic indicators in the group of 465 regions in the EU for the period 1980-1995 showed that leading rural regions:

- have both growth of industrial employment and services employment (Table 1.1);
- show a smaller decline in agricultural employment than lagging regions;
- show a population growth as well, whereas in lagging rural regions population growth stagnates;
- tend to have a lower unemployment rate than lagging regions.

However, for quite a number of socio-economic indicators hardly any differences have been found between leading and lagging rural regions like:

- participation rates;
- education level of the population;
- share of female employed in total employment;
- farm holders with other gainful activities; and
- small differences in the sectoral structure of employment.

Socio-economic characteristics of the case study regions

In order to make an in-depth analysis of the forces that hamper or encourage employment development in rural regions, 18 case studies in leading and lagging rural regions in nine EU member states were carried out (Fig. 1.1). Since Belgium has few rural regions, we have not selected a lagging rural region in this country. Instead, a French lagging region close to the Belgian border was selected. The selected rural regions are not unique in their development patterns or locations, but can provide insights with regard to the process of employment growth/stagnation and should generate lessons relevant for other rural regions too. Nevertheless, the selected regions reflect a wide range of characteristics with regard to their location, industrial tradition and physical structure.

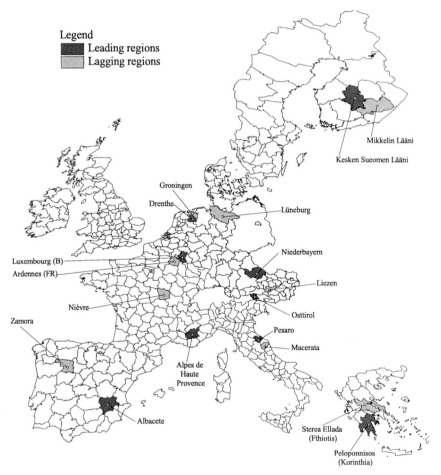

Figure 1.1   Selected case study regions in RUREMPLO
Source: LEI; RUREMPLO project.

Table 1.2 Distribution of employment over sectors (%) and employment growth in the case study regions (% p.a.)

	Year	Sectoral distribution employment			Period	Regional total employment growth	Difference employment growth region/ country [a]
		Agriculture	Industries	Services			
Leading regions:							
Luxembourg	1994	8	20	71	'80-'92	0.9	1.0
Niederbayern	1992	10	44	46	'80-'93	0.9	1.1
Korinthia	1991	33	22	45	'81-'91	0.8	1.3
Albacete	1995	12	30	58	'80-'95	0.2	0.9
Alpes de H. Prov.	1996	6	21	73	'81-'92	0.5	0.7
Pesaro	1995	5	43	52	'82-'95	-0.5	0.5
Drenthe	1995	7	27	63	'80-'91	3.6	1.3
Osttirol	1991	10	35	55	'81-'91	0.6	0.5
Keski-Suomen L.	1995	8	30	62	'80-'93	-1.2	0.2 [b]
Lagging regions:							
Lüneburg	1990	7	32	61	'80-'90	0.2	-0.5
Fthiotis	1991	34	20	46	'81-'91	-0.6	0.1 [b]
Zamora	1995	28	20	52	'80-'95	-2.2	-1.1
Ardennes	1990	8	37	55	'81-'92	-0.6	-0.7
Nièvre	1996	8	27	64	'81-'92	-0.6	-0.6
Macerata	1995	10	39	51	'82-'95	-1.5	-0.6
Groningen	1995	2	24	70	'80-'91	2.4	-0.3
Liezen	1991	10	34	56	'81-'91	-0.4	-0.8
Mikkelin L.	1995	16	26	58	'80-'93	-1.8	-0.2

Source: LEI, RUREMPLO project.
a) Measures the difference in percentage points of non-agricultural employment growth in the region and non-agricultural growth in the country; b) For country-specific reasons, we deviated for these regions from the general selection criteria.

The sectoral structure of employment is diverse: there are regions with a low and a high share of employment in agriculture (Table 1.2). It also appears, that regions with a high share of agricultural employment like, for example, Korinthia and Albacete, can experience a comparatively high employment growth in other economic sectors. On the other hand, there are also regions with a high share of agricultural employment that had a bad performance in non-agricultural employment growth. On the whole, it looks as if lagging regions tend to have a somewhat larger share of employment in agriculture, whereas leading regions tend to have a larger share of employment in industries and services.

Both leading and lagging case study regions were faced with a decline in agricultural employment and an increase in service employment since the 1980s. The most striking difference in employment development was the increase in industrial employment in a number of leading regions, which was absent in lagging regions (except for Zamora). Besides, employment in services increased in most of the leading regions at a higher rate than in the lagging regions. So most of the lagging case study regions follow the general trend in modern societies that employment in agriculture and industries decreases and that employment in services increases, whereas leading regions tend to deviate from this pattern.

Population and topography

The area size and population size in the case study regions reflect a wide range: the area size varies from 2,000 km² to 16,000 km² and the population size from 50,000 to 500,000 inhabitants. These differences in size are due to some country-specific differences in the size of the labour market area and due to pragmatic reasons for the selection of regions for case studies. The population density in the case study regions varies from 13 inhabitants/km² to 188 inhabitants/km², reflecting the fact that the case study regions cover both 'most rural regions' and 'intermediate regions'. Leading regions showed an increase in population during the last decade, whereas in the lagging regions population declined or showed only a moderate increase. This pattern reveals that employment growth tends to be accompanied by population growth.

In about half of the case study regions, the population is concentrated in one part of the region or in a few main centres (valleys) in the region. With the exception of Keski-Suomen Lääni, this is due to the mountainous state of the region. In the other case study regions, the population is dispersed across the region. It is striking that leading regions more often show a concentration of population relative to lagging regions. This can indicate that there is a relation between a concentration of actors and activities on the one hand, and employment growth on the other hand.

Striking issues in the pairs of leading and lagging rural regions

In order to go beyond the statistical analysis of employment development in the case study regions, we designed a conceptual model that can be used for a qualitative analysis of employment dynamics: the field of force of a rural region. This conceptual model is based on the mixed exogenous/endogenous approach, which considers rural development as a mesh of networks of local and external actors, in which resources are mobilized, and in which the control of the process consists of an interplay between local and external forces (see Chapter 2). Within the conceptual model, we have distinguished three closely related components in the rural region: local resources, economic activities and actors. The component of local resources refers to physical infrastructure (roads, railways, ports etc.), natural resources (like wood and hydropower) and rural amenities. Economic activities refer to all kind of activities in the agricultural, industrial and service sector. Actors are embodied with capacity (knowledge, skills and attitude) and interact with each other in networks. Moreover, actors can be involved in all kinds of relations with the outside world like the exchange of products, services and know how, contacts with policy makers outside the region, and migration.

This field of force offers a framework in which factors encouraging and hampering the maintenance or increase of employment opportunities can be revealed. It was used for making a template with questions for the case study analysis. This template was used in all case studies, which guarantees their mutual comparability. Below we highlight the most striking issues in the field of force of each pair of leading and lagging case study regions.

The case studies at the Belgium border: Luxembourg (B) and Les Ardennes

Both Luxembourg (Belgium) and the neighbouring Les Ardennes (France) were confronted with a restructuring of their declining iron and steel industry in the 1970s. In the 1980s Luxembourg (B) recovered from this crisis, whereas the employment development in Les Ardennes still stagnated (see Chapter 4). Multiple factors made employment growth easier in Luxembourg (B) than in Les Ardennes. A first factor was a sector mix already largely diversified into services in Luxembourg (B), which made restructuring therefore more manageable than in Les Ardennes. A second factor was the proximity of the growing economy of the neighbouring Grand Duché de Luxembourg, which stimulated immigration in Luxembourg (B) and hence population growth and the development of services. A third factor was a regional institutional framework concentrated among fewer municipalities and business supporting agencies, which made it easier for Luxembourg (B) to design and implement an effective territorial approach to employment growth. On the other hand, in Les Ardennes the large number of municipalities grouped into several associations and syndicates of different purposes was not favourable to design, agree upon and implement an effective strategy towards employment growth. As a fourth factor can be mentioned that the unified development and strategy developed by the political and economic leaders of Luxembourg (B) rested both on endogenous and exogenous factors. The local policy makers were clearly the initiators of an employment replacement strategy that included building and exploiting strong external networks. Through internal networks they realized an ambitious policy of economic and territorial development by ensuring a political and social consensus. Exogenous factors at work were settlement of foreign enterprises or subsidiary business and public and private investment coming from the outside and the improvement of the road, highway and telecommunication networks which opened the region. In contrast, Les Ardennes clearly suffered from a lack of leadership and unified strategy due to the economic heterogeneity of the region and lack of solidarity between the different parts of the region. Forces exogenous to the region mainly explain the regional development and employment dynamics. Priorities of policymakers in regional economic policy were the recovering of industrial competitiveness rather than employment. Entrepreneurs tended to focus on the manufacturing industry, not on the service sector.

According to the authors, a main lesson for employment creation in rural regions is the strengthening of the capacity of local actors to play an active role in developing and implementing together a common employment strategy. This capacity is reflected in the way in which local actors in Luxembourg (B) cooperated not only with each other but also with actors outside the region, in order to design and implement specific projects in line with the strengths and opportunities of their region. In addition to this first lesson, strengthening internal as well as external networks is another key lesson. Internal networks in Luxembourg (B) were stronger than in Les Ardennes because of the attitude of many local leaders, which was reinforced by a common sense of solidarity towards specific employment challenges and a centralization of regional institutions.

The German case studies: Niederbayern and Lüneburg

For more than four decades, the German case study regions were situated next to the 'iron curtain'. Despite this peripheral location, Niederbayern showed a very dynamic development: its employment and population growth were well above the national average during the period 1980-1995. On the other hand, the lagging region of Lüneburg faced a stagnation of employment and population in the 1980s. However, after the German unification in 1990 and the opening of the eastern border, employment and population dynamics have improved in Lüneburg, in particular due to a boom in the construction sector (see Chapter 5).

A crucial factor for the development success of Niederbayern has been its strong regional identity, a common perception shared by a great majority of regional actors in business as well as public administration. With respect to economic development, this common view is probably more rooted in a consciousness of (past) regional disadvantage than in a shared vision of the future. The common view finds its expression in effective territorial institutional structures and networks. This territorial cooperation enables regional actors to deliver a strong coherent message in lobbying for investment and support from outside. On the other hand, Lüneburg lacks a coherent regional vision, and local layers (at *Kreise* level) tend to compete with each other in the process of employment development. Another main difference between Niederbayern and Lüneburg deals with their position in the larger regional administrative unit to which they belong. Niederbayern is part of the *Bundesland* Bayern, which applied a strategy of decentralization of economic activities. Hence Niederbayern was encouraged to follow a territorial economic development approach, based on indicative regional development plans. These plans were launched and coordinated by the *Bundesland* Bayern and focused not exclusively on regional centres, but supported small and medium sized enterprises (SME) and the large number of pluriactive farm families as well. In addition, improvements in the transport infrastructure, the university and higher technical schools, which were set up at the initiative of Bayern in the 1970s, fostered economic development and technology transfer in the region. On the contrary, Lüneburg could not benefit from a decentralization of economic activities out of the main economic centre, simply because the *Bundesland* Niedersachsen does not have a main urban agglomeration like München in Bayern. Neither could Lüneburg benefit from an outflow of economic activities from the neighbouring Hamburg, as this city applied a concentration strategy of investments within its urban perimeter.

The German case studies prove that the chances for positive development performance are better if the territorial perimeter of the relevant public and private institutions involved in the regional/rural development process, be it public administrations, labour offices, chambers of commerce, or other non-governmental organizations, match the same territory. This helps by reinforcing regional identity, creating a commonly shared development vision, cooperating and coordinating joint development efforts. Where such a territorial match is achieved, it seems easier to set up networks of partners, strengthen coherence and cohesion internally and mobilise external support from outside the region.

The Greek case studies: Korinthia and Fthiotis

Both regions have a quite similar employment structure: at the beginning of the 1980s, nearly half of the labour force was employed in agriculture; at the early 1990s, this share was reduced to one-third (see Chapter 6). The high dependence on agriculture and the large agricultural exodus put an enormous pressure on the regional production system to generate jobs in the non-agricultural sectors. Korinthia, the leading region, managed to absorb the labour exodus from agriculture, mainly by an increase in employment in services, whereas Fthiotis, the lagging region, was not only faced with a large agricultural exodus, but with a decline in industrial employment as well. Employment growth in its services sector was insufficient to compensate for the losses in the agricultural and industrial sector.

Korinthia and Fthiotis show some main differences in their agricultural and industrial sectors. In Korinthia, agriculture is more intensive and provides higher agricultural incomes compared with Fthiotis. This slowed down the exodus of labour from the agricultural sector. Due to its proximity to Athens, Korinthia already experienced industrialization in the 1960s. In Fthiotis, industrial growth started in the 1970s. As a result of the EU membership in 1981, competitive conditions became harder. Korinthia had a more favourable industrial mix to face these new conditions and managed to stabilize its industrial employment, whereas industrial employment decreased in Fthiotis.

Networks are weakly developed in both regions, suffering from little interaction among actors and lack of cooperation between sectors. Despite this general weakness, the authors perceive some relative differences. This especially applies with regard to the external networks of Korinthia, which were less weak than those in Fthiotis, due to a more open attitude and a higher response to external changes of its actors.

Key lessons for employment creation from the Greek case studies refer to the need to strengthen the capacity of local actors, to develop internal networks and to establish effective links with external actors.

The Spanish case studies: Albacete and Zamora

Until the mid-1970s, both Albacete and Zamora experienced a considerable out-migration, resulting in a population decline. With the creation of the Autonomous Communities in 1978, Albacete transformed from a poor, marginal region into a main centre in the new region of Castilla-La Mancha, along the dynamic Mediterranean Arc. On the other hand, Zamora remained at a disadvantaged location in the weakly industrialized region of Castilla y Leon. Albacete became a leading region in the 1980s and early 1990s, showing an increase in both industrial and service employment and a reversal of migration, whereas Zamora lagged behind in employment creation and continued to lose population through out-migration (see Chapter 7).

In the opinion of the authors, the main fact behind the success of Albacete is the capacity of its local actors to diagnose the socio-economic situation correctly, to define objectives, to make good use of structural policies and regional incentives and to coordinate their activities. This resulted in a positive atmosphere, large flows of development funding and business investments, new infrastructure and business, encouragement of product diversification and many pubic services. In contrast, the cautious and risk averse disposition of policymakers and entrepreneurs in Zamora constrained economic growth. Its political parties are dominated by powerful local figures, and wrangles between such individuals can tend to hamper the working of the local government. Industry and commerce entrepreneurs have different organizations between which there exists significant conflict. Entrepreneurial activity in Zamora originates locally and is limited to local markets, so that there are no meaningful connections outside the province. It can be said that Zamora is an example of socio-economic isolation with a weak development of internal and external networks.

The French case studies: Les Alpes de Haute Provence and La Nièvre

Les Alpes de Haute Provence (AHP) and Nièvre are both mountainous regions with valleys where industrial activities are concentrated, plateaux where agricultural activities dominate, along with many small towns, villages, rivers and lakes (see Chapter 8). In AHP, industry has never developed significantly because of the region's isolation and relatively remote geographical location. Due to the increase in the demand for leisure activities since the 1960s, AHP has entered the dynamics of tourism through the enhancement of many natural and cultural resources and winter sports holidays. In the 1980s, AHP experienced a relatively high increase in its service sector and was therefore labelled as leading region.

In contrast to AHP, Nièvre has a long industrial tradition in ironworks. It looks as if Nièvre has a favourable location close to Paris. On a more detailed look, it seems evident that part of the department remains rather isolated notably because the Morvan mountains are a true obstacle, especially in winter. Due to the closeness of Paris, Nièvre has a long tradition of seasonal emigration of agricultural and forestry workers or long-term out-migration towards the Parisian labour market. During the last centuries, Nièvre has supplied Paris with firewood. Nièvre has experienced an industrial crisis since the beginning of the 1980s, resulting in a huge decline in employment. Tourism is not so well developed as in AHP, due to access difficulties in wintertime. So, despite the growth of employment in the services sector in the 1980s, this was not sufficient to compensate for the losses in industry. Hence Nièvre was labelled as lagging region.

Like in other case studies, here again a main difference in the attitude of local actors can be perceived. In AHP, the beautiful diversified landscape associated with a Mediterranean climate not only attracts tourists, but highly educated people, setting up enterprises, unemployed people, looking for a cheap place to live, retired people and seasonal labourers working in tourism and fruit harvesting as well. This immigration results in a positive atmosphere in AHP, which brings about an active attitude of actors, who face and tackle innumerable challenges in the field of natural conditions, infrastructures, access, diversification of activities, attractive tourist and industrial sites, and the maintenance of services like shops, public services and local trade in small rural towns. Policy makers are able to identify the needs in the region and to deliver policies accordingly. They also have good contacts with the upper level authorities. Entrepreneurs (both local and in particular newcomers) are innovative in searching for new initiatives and products as well as new ways of marketing of products. On the other hand, Nièvre has a negative migration balance made up of emigration of young people under 30 years and an immigration of retirees. The continuous emigration of young workers erodes the human resource base and affects the atmosphere negatively. The attitude of the actors can be characterized as rather passive, waiting for jobs to be established by external forces. Most entrepreneurs are risk averse and not very innovative: they prefer to exploit the first steps in the production process and thereby miss many chances for value added. Policy makers tend to be defensive and less forward-looking and not able to establish good working contacts with the regional layers in Dijon.

The Italian case studies: Pesaro and Macerata

Pesaro and Macerata are located in the so-called 'third Italy', which experienced a spontaneous industrialization process characterized by small family firms and concentration in industrial districts during the last decades (see Chapter 9). Both regions are quite similar in their socio-economic characteristics: since the 1950s economic activities and population tended to concentrate in the coastal areas, resulting in a marginalization of the inner rural areas and an urbanization of the coastal areas. The industrial sector has a share of about 40% in both regions. The main and traditional industrial specialities are wood-furniture in Pesaro and footwear in Macerata, around which districts emerged.

In the years 1982-1995, Pesaro and Macerata were both faced with a decline in agricultural and services employment. The development of industrial employment made the difference: it increased in Pesaro and decreased in Macerata. Hence, Pesaro was labelled as leading region and Macerata as lagging region. Although the industrialization process has reached a mature stage in both provinces in the 1980s, this does not imply that it has achieved long-term stability and sustainability. On the contrary, maturity means that the previous stage of industrial growth built upon the reinforcement of the industrial districts has turned into a continuous cycle of crises and restructuring driven by global competitive pressure. The local industrial system reacts by shifting towards new segments of global markets, a process which requires new technologies, new specialities, new markets, and new local leaders and hierarchies. The outcome of this cycle is highly unpredictable: it may give rise to de-industrialization, decline or further success.

In the transformation and reorganization of the furniture district in Pesaro and the footwear district in Macerata in the period in question, firms in Pesaro showed a greater capacity in finding new markets, new niches and specialization than firms in Macerata. From this it appears that the regenerating capacity of actors is better developed in Pesaro. On average, firms are smaller in Pesaro, and individualism and willingness to take risks are more important than in Macerata; therefore, local actors show a greater capacity to react promptly to changing external conditions. Moreover, sector specialization also makes a difference. Wood-furniture production is much more complex in terms of technologically integrated production phases than footwear. On the one hand, this makes local actors able to specialize in many different 'segments' of the whole process, thus creating new firms and eventually new specializations. On the other hand, this complexity makes the local system much more sound and resistant with respect to global competition.

Emerging key messages from the Italian case studies are that industrial success based upon small-medium highly specialized enterprises is forced into a continuous process of reorganization and technological innovation, in which the capacity to move toward new markets, new niches, new technologies and specialization is strategically important. Further, the industrial crisis in Macerata reveals that employment policies directed at the agricultural and services sectors are needed to avoid dramatic impacts of industrial crises on the regional economy.

The Dutch case studies: Drenthe and Groningen

The Netherlands is a small and densely populated country with only five rural regions. Two neighbouring regions in the north of the country have been selected for case studies: Drenthe as a leading region and Groningen as a lagging one (see Chapter 10). However, these labels only apply for employment growth in the 1980s; if the period 1984-1996 is taken into account, total employment growth in Groningen exceeds that in Drenthe. This shows that 'leading' and 'lagging' is not a permanent situation, but depends on the period considered.

Due to their peripheral location in the country, Drenthe and Groningen have much in common. They both benefit from the outflow of firms from the congested economic centre in the western part of the country, although Drenthe benefits more as it is located closer to this centre. Internal networks in both regions are small, surveyable and characterized by easy communication. On the whole, networks are insufficiently directed towards actors outside their own region. An improvement in this sense is the strengthening of the cooperation of policy makers in the three northern provinces in recent years. Both regions profit from the city of Groningen, the main economic centre in the north of the Netherlands, which is located a few kilometres from the Drenth border. The university is one of the largest employers in the city. This has contributed to the creation of a knowledge infrastructure around the city of Groningen, which attracts many high-tech activities. The city of Groningen boosts employment in the northern part of Drenthe, as many companies look for settlement opportunities in the neighbourhood of this city.

The main differences between Drenthe and Groningen refer to the landscape and the distribution of cities in the region. Drenthe has a beautiful varied landscape. Traditionally, it was the main 'holiday province' in the Netherlands. In contrast to Drenthe, Groningen has a flat landscape, which attracts few tourists. In Drenthe cities are well spread over the region, whereas in Groningen the larger cities are located in a zone running from southwest to northeast. For a long time policy makers in Drenthe have applied a strategy towards a concentration of economic activities in well-defined zones around the major cities. Advantages of such a concentration are a relatively high density of enterprises, which stimulates the interaction among enterprises and which attracts other enterprises. This leaves space for other functions like housing, recreation, agriculture and nature in other parts of the province. The concentration of firms in Groningen in the southwest-northeast zone is more the result of infrastructural endowments and market forces.

Lessons for employment creation from these case studies refer among others to the concentration of economic activities in well-defined zones and cooperation with neighbouring regions in order to create critical mass.

The Austrian case studies: Osttirol and Liezen

Osttirol and Liezen are both alpine regions with settlement in the valleys (see Chapter 11). Whereas Osttirol shows a concentration of economic activities and population into the centre of Lienz, Liezen has a more scattered pattern. Until the construction of the Felbertauern road in 1968, Osttirol was quite isolated from the rest of the country. Since then, industrialization with both large firms and small and medium sized firms has taken place, with increasing employment in the industrial and service sectors. Hence, Osttirol was a leading region in the period under study. In contrast to Osttirol, Liezen has a long industrial tradition. Closures of big manufacturing firms in the 1980s resulted in a rapid decline of industrial employment, which was only partly compensated for by an increase in service employment.

In Osttirol local politicians played an important role in attracting new firms from outside and showed a cooperative attitude to firms. Active communities created good preconditions for industrial development (industrial zones with necessary infrastructure) and attracted firms with cheap land and cheap connection to energy, water and canalization, and generous support in the start-up period. Besides, firms were also attracted by the well-educated labour force with medium-level technical skills and keenness for education and permanent training. Another advantage of the labour force in Osttirol was its relative low labour costs.

At the outset of the 1980s, the industrial structure of Liezen was characterized by a few large industrial firms. Declining prices on world markets would have required strong measures of rationalization and a continuous change towards new products. However, this did not happen due to bad management, inflexible labour contracts, and political pressure. The relatively high industrial wages eroded incentives for self-employment, so entrepreneurial potential in the region was low. Except for the eastern part of Liezen, policy makers were not very active in attracting new firms. Instead they subsidized the declining industry. Furthermore, due to the lack of technical education facilities in Liezen, firms in Liezen experienced difficulties in finding qualified labour.

Osttirol and Liezen are districts that have a totally different position in their regional administrative context. Osttirol is part of the province of Tyrol, which has partly decentralized the decision process on the use of important regional funds. Due to good contacts between policy makers in Osttirol and Tyrol, Osttirol benefited considerably from these funds. Liezen is part of the province of Styria, which followed a strict method of central decisions on funds. Liezen has only a marginal position in this province, and coordination of local and provincial policy makers is insufficient. Hence Liezen hardly profits from regional funds.

Lessons for employment creation, which emerge from the Austrian case studies, refer to a cooperative behaviour of policy makers towards firms, to taking care of good contacts of local policy makers with higher administrative layers, and to the training of the labour force.

The Finnish case study regions: Keski-Suomen Lääni and Mikkelin Lääni

The selection of case studies in Finland was blurred by the deep economic recession in the early 1990s. So both the leading region Keski-Suomi and the lagging region Mikkeli lost employment in the period under study, which was made up of a decline in agricultural and industrial employment and a small rise in services employment (see Chapter 12). Nevertheless, industrial decline in Keski-Suomi was less than in Mikkeli, and, what is of high importance in the Finnish context, Keski-Suomi showed a population increase.

In Finland some 97.5% of the area comprises scattered settlement, and only 2.5% is occupied by cities or villages with densely built-up areas. During the last decades, rural population has tended to migrate from the outskirts of provinces to their capital areas and from all regions to the Helsinki region in the south. Because of this out-migration of rural areas, the story of forces behind employment dynamics in the Finnish rural regions is to a great extent a story about settlement structures and policies, and not so much about actors and networks.

In Keski-Suomi about 40% of the population lives in the capital town region around Jyväskylä. This is a dynamic regional centre with a university, which has a nation-wide positive image among students, which boosts the R&D infrastructure and supports the upkeep industrial know-how basis of the region. On the other hand, in Mikkeli only 22% of the population lives in the central area around the capital, and it lacks a catalyst for employment growth like the Jyväskylä University. Besides, the mix of industrial activities in the Jyväskylä area is export oriented and partly related to industries with favourable future perspectives, whereas the industrial structure in Mikkeli is less competitive. In the more remote parts of both regions, the economic structure is less diversified and predominantly directed to industrial and agricultural activities with declining employment opportunities, resulting in an unavoidable outflow of labour from these areas.

A key message for encouraging employment in low populated regions, emerging from the comparison of the Finnish case study regions, is the need for a strong regional centre. The concentration of economic activities and population gives rise to critical mass for becoming competitive and achieving synergies. If necessary, population from other parts of the country can be attracted by financial incentives to such regional centres.

Lessons for employment creation in rural regions

General guideline for employment creation

Since the socio-economic, physical and geographic situation of rural regions varies widely, there is no one unique development path towards more jobs. So the lessons formulated below should not be considered to be the 'success formula', which always results in more jobs. The lessons have to be seen as building blocks, which may contribute to shaping preconditions for employment creation under certain circumstances. Despite the multiple development trajectories, we give a general guideline for employment creation in rural regions, based on the experience in the case study regions:

- make a comprehensive territorial development plan, based on the strengths, weaknesses, opportunities and threats of the region, and integrate all measures and projects within the scope of this plan;
- improve the capacity (knowledge, skills and attitude) of local actors;
- strengthen the cooperation of local actors and the cooperation of actors inside and outside the region.

Within the framework of this guideline the following lessons – if suiting the needs of the region – may be selected and applied (see Chapter 13).

Lessons with regard to local resources

- integrate infrastructure investment in a broader development process;
- pay attention to distinct modes and technologies of infrastructure in rural regions;
- valorize rural amenities;
- improve the perception of amenities by rural actors.

Lessons with regard to economic activities

- follow a multisectoral approach;
- support the integration of agriculture in the rural economy;
- both specialization and diversification can be successful strategies;
- enhance facilities for new and small enterprises;
- focus on the local productive system;
- strengthen zoning of economic activities by spatial planning.

Lessons with regard to actors

- enhance capacity building of local actors;
- strengthen internal and external networks;
- attract newcomers;
- define the right labour market area;
- aim at the appropriate regional mix of skills;
- be aware of changes in labour demand by industrial firms;
- encourage part-time labour and self-employment.

Theoretical and Methodological Framework

Ida J. Terluin

Introduction

To analyse employment dynamics in rural Europe, two items have to be tackled. First, what is meant by rural Europe, and second, what are the driving forces behind employment dynamics. The first question is answered by a discussion of the different approaches to rurality and the second one by a review of regional economic growth theories and the debate on economic development in rural regions. The results reveal that the mixed exogenous/endogenous development approach is suitable for the analysis of employment dynamics in rural regions in Europe. Based on this approach, a field of force of a rural region is designed for analysing employment dynamics in case study regions.

Approaches to rurality

The debate about the conceptualization of rurality 'strikes right at the heart of rural studies' (Pratt, 1996:71). Among the multiplicity of meanings of rurality, two main approaches can be distinguished: concepts that perceive the rural as a distinctive type of locality and concepts that describe the rural as a social representation. Or in other words: 'the rural as space and the rural as representing space' (Halfacree, 1993:34). In the first approach, spatial classifications are made based on a variety of land use and/or socio-economic variables, often resulting in areas that can be mapped (Errington, 1994:367). Blanc (1997) has proposed a division of the first approach into a spatial category and a territorial category which gives the following classification of meanings of the rural:

(1a) Spatial approach

This approach is based on the idea that rural space has some characteristics which are different from other (usually urban) spaces. It has its roots in models of spatial economics like the central place theory of Christaller and Lösch, in which space is regarded as a set of points in which economic agents (producers, labourers and consumers) have to find out the optimal location (Blanc, 1997:1-3). Such models are often characterized by a hierarchical vision on space, a ranking of goods and services, and the presence of agglomeration and dispersion forces. The starting point in the structuring of space in such models is a concentration of economic activities in a centre, due to economies of scale and transport costs. This concentration of firms and labourers attracts retail business and personal services, which results in a diverse supply of products, skills and information. This diversification can be seen as a positive externality of the centre and is an incentive for a new round of settlement of firms and labourers in the centre. This agglomeration is a cumulative process. However, dispersion forces are also at work. Competition for land use and negative externalities like pollution and congestion will push activities and agents out of the centre. Activities with a less intensive land use like agriculture, and people who enjoy the benefits of the centre less, move away. The complex interplay of agglomeration and dispersion forces results in a structuring of space where centres and peripheries differ not only in the density of jobs and people, but also in the structure of economic activities and characteristics of households. In this way, a functional specialization of spaces emerges. The further away from the centre, the more likely one is to find less intensive land-use activities, basic services (which are frequently consumed and which do not need to be produced on a large scale to become profitable), and a composition of the population which consists of a relatively high proportion of non-active people. The periphery is often associated with traditional economic and social structures, and expressed in terms of a dependent relationship to the urban centre. In the spatial approach, rural coincides with periphery.

Descriptions or indices of rurality, which appear within the spatial approach, usually perceive rural areas as those areas satisfying certain statistical thresholds in socio-economic or socio-cultural dimensions (see for example Halfacree, 1993:24-5, Pratt, 1996:70, Borgstein *et al.*, 1997:14-18). These thresholds may refer amongst others to land use (like agriculture, forest and nature), share of agriculture in employment, population density, built-up area, frequency of contacts, density of societies and crime rates.

(1b)   Territorial approach

This approach abandons the strong interweave of rural with agriculture and the urban-rural dichotomy in the spatial approach and emphasizes the economic diversification of rural areas. According to the territorial approach (sometimes also referred to as local economy approach; see Saraceno, 1994:456), space is divided into territorial entities which cover a local or regional economy. Each territorial entity includes both agricultural, industrial and service activities and consists of one or more centres and open space. Some of these territorial entities are densely populated, have a metropolitan centre and a small area of open space, whereas other territorial entities are characterized by a low population density, one or more small or medium sized towns, and a large area of open space. The first territories can be labelled as 'urban territories' and the second as 'rural territories'. In between these two opposites, a wide range of different configurations exists. The relationship between the different territories should not be conceived in traditional terms of dependence of one territory on another, or core-periphery dynamics, but rather in terms of a set of non-hierarchical competing local economies in the world market (Saraceno, 1994:469).

The concepts of territory, local and regional economy used in this approach are rather vague: they refer to spatial entities in which economic, social, political and cultural life is structured and whose size may vary (Keating, 1998). Besides, although rural is related to low population density, no strict criteria are given. This implies that the territorial approach enables the coexistence of different rural territories, depending on the definition of the user.

(2) Constructivist approach

According to this approach, space appears as a social representation: a mental construct which guides us to deal with the complexity of the social world. Or in other words: rural is an image in our mind, which helps us to prescribe and organize our behaviour and responses (Halfacree, 1993:29). The social representation of a specific actor is an amalgam of personal experiences and handed-down beliefs, propagated through literature, the media, the state, family, friends and institutions (Halfacree, 1993:33). Hence the mental constructs of rural vary among actors, depending on place, time and social group, and may include such concepts as countryside idyll, agriculture, quietness, non-urban, nature, recreation, open space, backwardness etc. In some of these social representations, rural is conceived as a production asset and in others as a consumption asset (Hoggart *et al.*, 1995:28).

Actors involved in the construction of representations of rural are diverse: they can, for instance, be permanent or temporary users of space, policy makers or academics. Conflicts and controversies about alternative uses of rural space indicate that social representations are related to the issue of power (Pratt, 1996:70). Such conflicts may arise, for instance, between farmers' interest groups, which want to exploit space and therefore adapt some landscape elements, and tourist or nature conservation interest groups which prefer to maintain cultural and natural values of rural space.

Choice for territorial approach

The concept of rural that one uses, depends on the problem to be solved. In a study focused on employment development in rural Europe, rural is explicitly linked with a distinctive type of locality. The territorial approach is appropriate, as it enables us to consider rural space as a territorial entity, which covers a regional economy with agricultural, industrial and services activities. By doing so we can take all relevant economic relationships into account. Our choice implies that we distinguish a large number of rural regional economies in Europe, which are briefly labelled as 'rural regions'. Usually, the size of our regions reflects that of a labour market area.

Theoretical approach to economic development of rural regions

In this section we discuss how employment dynamics in rural regions have theoretically been conceptualized. The development of employment is closely related to the development of the production structure of the regional economy. Therefore, we can use theories on regional economic growth for the theoretical conceptualization of employment dynamics. Below, we firstly focus on a general review of regional economic growth theories, disregarding whether such theories can be applied in rural regions. As a next step, we turn to the debate on economic development in rural regions.

Review of regional economic growth theories

Camagni (1992:1) describes regional development as: 'the ability of each region to produce with a comparative advantage the goods and services that are demanded by the national and international system of which they form a part'. This description emphasizes that the process of regional economic change is part of interregional interdependencies. In this process some regions experience a more favourable development than others since competitive economic struggles 'cannot have winners without also having losers, although the absolute level of regional development may rise over a long historical period' (Healy and Ilbery, 1990:298). Regional firms and production circumstances are at the core of the economic development process as firms' ability to adapt to changes is decisive for realizing economic growth (Lambooy *et al.*, 1997:73). Theories of regional economic growth can be divided into four groups, depending on the factors in the production function: traditional models, pure agglomeration models, local milieu models and territorial innovation models (Fig. 2.1). The sequence of these models is such that the factors in the production function increase in complexity. Besides, the models reflect a certain degree of chronological sequence: the traditional models were prevalent in the 1950s, the pure agglomeration models in the 1960s, the local milieu models in the 1970s, and the territorial models have dominated since the 1980s. The four groups are briefly discussed below[1].

Traditional approach

In the traditional approach, output is assumed to be a function from the input of labour and capital. The neoclassical growth theory and the Keynesian export base theory are the main exponents of this approach. In the neoclassical growth theory, the evolution of regional disparities depends on the availability and the interregional mobility of production factors. Flexible prices and wages on regional markets guarantee the full utilization of regional resources. Given identical production functions, capital tends to move to regions where labour is abundant and cheap while labour will move in the opposite direction. These flows

---

[1] Unless otherwise indicated, this discussion is based on Molle and Cappelin (1988), Healy and Ilbery (1990), Malecki (1991), Camagni (1992), Lambooy *et al.* (1997) and Rijswick (1997).

	Production function [a)	Theories
Traditional models	$Y = f(L, K)$	Neoclassical growth theory; Keynesian approach: export base theory.
Pure agglomeration models	$Y = f(AE, L, K)$	Cumulative causation theories; Growth pole theories.
Local milieu models	$Y = f(LM, L, K)$	Endogenous growth models; Theories based on changes in the organization of labour.
Territorial innovation models	$Y = f(I, LM, L, K)$	Incubator theories; Product life cycles; Innovative milieu; Global-local paradox; Porter's theory on the competitive advantage of nations; Storper's theory on the region as nexus of untraded interdependencies.

Figure 2.1   Classification of theories on regional economic growth
a) Y: income or output; L: labour; K: capital; AE: agglomeration effects, due to external effects or scale economies; LM: local milieu, which includes items like space, networks, trust, culture and policies; I: innovation.

continue until returns to capital and wages for labour are equal in each region. So in the end, per capita incomes will converge. The export base theory divides economic activities into basic activities for export and non-basic activities for internal consumption. According to this theory, regional economic change depends on the proportion of the economic activities in a region that produce goods or services for export. A growth of the basic activities enlarges the flows of money into the region, increases the demand for goods and services within it, and causes a corresponding increase in the amount of non-basic activities. The size of this multiplier effect depends on the amount of money that is spent in the region.

Pure agglomeration models

Due to a concentration of labour and capital in a specific location, external effects or scale economies may arise. We discuss here two exponents of this approach: cumulative causation theories and growth pole theories. The main idea behind cumulative causation theories is that once disparities come into existence, a self-reinforcing process starts that, in the absence of catastrophic events, maintains the status of growing areas. Contrary to the convergence trend in the neoclassical theories, here divergence among regions is the result. The most well known exponent of these theories is the Swedish Nobel price winner Gunnar Myrdal. His theory assumes that firms in wealthy regions have several advantages relative to firms in lagging regions: the size of the market is larger, which enables scale economies; the education level of the labour force is higher; and possibilities for innovations are more favourable. An expansion of production in wealthy regions results in a migration of (often high skilled) labour from lagging regions. The one expansion in the wealthy region induces another expansion as new firms are attracted by the already existing concentration of economic activities. The production of consumer services will also expand by the rising population in the wealthy region. The increase in tax revenues enables the provision of infrastructure. This cumulative process of concentration and expansion of economic activities in the wealthy region implies that lagging regions are deprived from labour and capital, the so-called 'back wash effects'. Furthermore, the non-expansionary regions have increasing disadvantages since these regions cannot afford to keep up a good infrastructure, a good school system and other public utilities. This will again increase their competitive disadvantages. Moreover, the entire system of valuations of people living in backward regions will change and influence further development negatively. On the other hand, in the course of time a deconcentration of economic activities out of the wealthy region starts due to high land prices, shortages at the labour market and traffic congestion, the so-called 'spread effects'. However, these spread effects cannot undo the divergence in economic growth in wealthy and lagging regions.

The basic idea of the growth pole theories is the existence of a leading or propulsive firm, which acts as a growth pole and which can stimulate other industries and businesses through multiplier effects. Leading industries are characterized by their newness, high technology and strong linkages with other sectors, while propulsive industries can be seen as relatively large firms, belonging to a growth sector and having a high ability to innovate and to generate growth. The multiplier or polarization effects are threefold:

- technical multipliers by linkages with upstream and downstream industries;
- income multipliers as a result of increased employment, which induces the demand for consumer services;
- psychological effects as the establishment of a large firm may create an optimistic atmosphere.

These multipliers may enhance a favourable regional economic environment around the leading or propulsive firm, which can be referred to as 'growth pool'.

Local milieu models

In this group of theories, various factors in the local milieu such as skills of the labour force, technical and organizational know-how, and social and institutional structures, affect the revenues from the input of capital and labour. Within this group, a distinction can be made between endogenous growth models and theories based on changes in the organization of labour. The endogenous growth models usually refer to agglomerated but non-metropolitan areas with small and medium sized firms. These local economies are characterized by entrepreneurship, production flexibility, district economies, and some collective agents, which act as a catalyst in the development process. Many applications exist, such as the development-from-below-approach and industrial districts.

The starting point in the theories based on changes in the organization of labour is that the composition of the labour force in terms of skills, costs, mobility, number etc. varies between regions. These differences in the labour force affect the location decision of firms. We discuss here the theory on the spatial division of labour and the regulationist theory. The theory on the spatial division of labour assumes that the spatial inequality is both produced and used by firms in their search for favourable conditions for profitable production, and that there are rounds of investment and disinvestment. So investment is attracted to areas where there are profitable opportunities, and disinvestment occurs in areas where profitable opportunities are exhausted. The pattern of geographical differentiation is continuously transformed by new (dis-)investment rounds. In these theories, the development of a particular region results from the interaction of external factors (the national and international context) and local actors (available material resources and factors of production, industrial structure and social composition of the area).

Regulationist theories suggest that capitalist economies develop through a series of regimes of accumulation: the way in which labour is organized and controlled in the production process like Fordism and post-Fordism. The transition periods from one regime to another are of critical importance as these are accompanied by a decline of the industrialized regions of the previous regime and the emergence of newly industrialized regions under the next regime. According to the proponents of this approach, capitalist economies have been in a transition phase between Fordism and post-Fordism since the mid-1970s.

Territorial innovation models

These theories are an extension of the local milieu models in the sense that they add the diffusion of innovations, which is considered to be the engine behind growth. This implies that technological ability to adapt to innovations is crucial for entering into new types of production and new markets. Various theories can be classified in this group, such as incubator theories, product life cycle theories, innovative milieux, theories on the global-local paradox, Porter's theory on the competitive advantage of nations, and Storper's theory on the region as nexus of untraded interdependencies.

Incubator theories stress the natural tendency of R&D and innovation activities towards areas with a concentration of people and activities. These areas can benefit from external economies, spin-off effects of a skilled labour force, and organizational and technological know-how, which create a fertile environment for innovations. Together with the emphasis on innovation as driving force behind economic growth, the implication of the incubator theories is that present economic core areas will also be the core areas of tomorrow. So core regions are characterized by continuity.

The product life cycle theory builds upon the incubator theories. The product life cycle can be divided into three stages: innovation, growth and maturity. The central idea in the product life cycle theory is that locational shifts occur in the various stages of the cycle. The innovation phase takes place in areas with a concentration of technical and scientific labour, whereas the maturity phase of the product requires areas with large amounts of low-cost labour such as peripheral regions.

The theory of innovative milieu can be considered as a dynamic counterpart of the industrial district (Camagni, 1995a, b). It conceptualizes the innovation-driven industrial behaviour within a geographical area and connects the local milieu (i.e. local production fabric of flexible small and medium sized enterprises environment) with innovation processes. Such processes provide dynamic efficiency to the local milieu and are reflected in the capacity to imitate and create technology, fast reaction capability, capacities for shifting resources from declining production sectors to new ones while utilizing the same fundamental know-how, and the capacity to regenerate and restructure a local economy hit by external turbulence. The continuing reproduction of the innovation capability of the innovative milieu may by no means only be attributed to its internal functioning. External energy in the form of technological, organizational or market information is crucially needed. This information is obtained by means of trans-territorial networks. So two types of networks can be distinguished in the innovative milieu:

(1) local networks, in which the element of proximity – spatial, cultural or psychological – generates three distinctive features: density of relationships, informality and openness;
(2) trans-territorial networks, which are systems of relationships for long distance, where the non-proximity of partners implies and requires relatively few links, greater formalization of relationships, network selectivity and closure.

The innovative milieu can be positioned within the global-local paradox, in which globalization and regionalization simultaneously occur: on the one hand, there are firms, which make the different parts of their product in different parts of the world and which sell their products all over the world, and on the other hand, firms which concentrate in regional clusters. These different organization forms of firms are studied in the scope of the 'network approach' (Capello, 1996) by various schools from social sciences and organization theory. Transaction costs (costs of using the market), complementarity and trust are main items in the network approach. In the scope of our discussion of regional economic growth, it can be stated that networks are an important local milieu factor.

Porter's theory (1990) on the competitive advantage of nations addresses the key question: 'why do firms based in particular nations achieve international success in distinct segments and industries?'. Porter assumes that firms can and do choose strategies that differ. The home nation of a successful international firm is that in which the essential competitive advantages of the enterprise are created and sustained. Porter distinguishes two basic types of competitive advantage for firms: lower costs and product differentiation. Firms can gain and sustain international competitive advantage through improvement, innovation and upgrading. Porter uses a 'diamond' to explain the determinants of national advantage. He distinguishes four main variables, which individually and as a system, create the context in which a nation's firms are born and compete. These four variables are as follows:

- factor conditions;
- demand conditions;
- related and supporting industries;
- firm strategy, structure and rivalry.

In addition to these four main variables, Porter distinguishes two other important variables:

- chance;
- government.

These six variables form together a dynamic mutually reinforcing system of the determinants of national advantage: the so-called diamond. The determinants themselves are influenced by cultural, social and political factors. The diamond is transformed into a system by two elements: domestic rivalry and geographic concentration. Domestic rivalry stimulates the upgrading of the whole diamond, and geographic concentration strengthens the interactions within the diamond. Although Porter's analysis is carried out at a national level, it 'can be readily applied to political or geographic units smaller than a nation' (Porter, 1990:29). Porter points that successful firms are frequently concentrated in particular regions within a nation.

Storper's theory (1995) on the region as nexus of untraded interdependencies deals with the question of why regions keep emerging as centres for new rounds of economic growth in an age of increasing ease in transportation and communication. Firms are tied to other firms through formal exchanges (i.e. the input-output linkages) and through untraded interdependencies. These include labour markets, public institutions, and conventions like rules for action, customs, understandings and values. The untraded interdependencies can also be seen in terms of 'regional production culture' or 'civic culture', i.e. the set of virtuous connections of economic coordination, which mobilize capacities for efficient economic action. They form the public assets of the production system. According to Storper, untraded interdependencies are necessary in the capitalist system, and these interdependencies are located in the region. The untraded interdependencies may differ among regions. All production systems are subject to uncertainty: between producers, between producers and labourers, and between producers and consumers. These uncertainties are mainly solved through conventions, which are taken-for-granted rules and routines between the partners in different kind of relations of uncertainties. There are different combinations of uncertainty and different conventions among regions, resulting in different 'frameworks of economic action' or different 'worlds of production'. Some of these worlds of production are more competitive than others like, for example, Silicon Valley. The evolution of the production system is strongly dependent on its underlying conventions. These affect the labour market, the input-output system and the knowledge system, and tend to push the production system from generality into specificity. This evolution is path dependent in the sense that it involves interdependencies between the choices made over time and that it is irreversible.

Debate on economic development in rural regions

From this review of regional economic growth theories, we now move to the debate on economic development of rural regions. This debate is on the one hand concerned with theories on economic growth in rural regions, and on the other hand with the question how rural development policy can stimulate economic growth in rural regions. In this debate, three approaches can be distinguished:

(1) the exogenous development approach;
(2) the endogenous development approach;
(3) a mix of the exogenous and the endogenous development approach.

These approaches reflect more or less a chronological sequence of conceptualizing rural development. They show close correspondence with successively the agglomeration models, the local milieu models and the territorial innovation models. The concepts have different implications for the strategies of local actors and for rural development policies. The three concepts are discussed below.

The exogenous development approach

Main elements of exogenous models are that rural development is considered as being transplanted into particular regions and externally determined, that benefits of development tend to be exported from the region, and that local values tend to be trampled (Slee, 1994:184). Exogenous models are based upon a view that modernization results in a division of economic activities between urban and rural: urban areas become the domain of industries and services and rural areas that of agriculture. The agricultural sector covers several functions in this system: it provides food for the urban areas, it is a source of purchasing power for commodities of the industrial sector, a source of capital and labour for the industrial sector, and a source of foreign earnings to support the development process of the urban areas. Since these functions reflect a dependency of agriculture on the urban sector, the process of agricultural development and hence rural development is seen as dependent on and exogenously determined by the urban sector.

Till the 1970s this was the dominant model for explaining rural development. In the European countries, it was reflected in a rural development policy directed towards modernization of the agricultural sector; as this proved insufficient to stabilize the rural economy, a policy of branch plant – derived from the growth pole theory – was also adopted, in which manufacturing firms from urban areas were encouraged to move into rural areas in order to create employment opportunities for the rural population. By the late 1970s these policies fell into disrepute since they did not result in sustainable economic development of rural regions (Lowe *et al.*, 1995:89-91).

The endogenous development approach

Endogenous development is to be understood as local development, produced mainly by local impulses and grounded largely on local resources (Picchi, 1994:195). In contrast to the exogenous model, the benefits of development tend to be retained in the local economy and local values are respected (Slee, 1994:184). Within rural policies the emphasis shifted towards rural diversification, bottom-up approach, support for local business, encouragement of local initiatives and local enterprises, and provision of suitable training (Lowe *et al.*, 1995:91).

This approach is closely related to the local milieu models, such as the endogenous growth and industrial district models, in which the institutional context of the economic activities plays an important role. An industrial district can be seen as 'a local thickness of inter-industrial relations which is durable in time and forms an inextricable network of positive and negative externalities (and) historical-cultural inheritances' (Becattini (1987), quoted in Iacoponi *et al.*, 1995:34-5). In this system, an agglomeration of small and medium sized firms exchanges semi-finished products, which can be described as a collective production process. In this process transaction costs are very low. Technology employed in each firm is very similar and well known to everyone due to a local technological atmosphere. Hence, information costs are also very low. Relations between firms and persons in the local system are not only established by national regulations, but to a large extent by local regulations, rules and customs which have their roots in local historical culture (Iacoponi *et al.*, 1995:34).

Two specific 'rural' theories within this approach can be put forward: the community-led rural development theory and Bryden's theory on the potentials of immobile resources for creating competitive advantages in rural areas. These are briefly discussed below.

The community-led rural development theory focuses on the strengthening of the self-help capacity of local actors, which is considered to be a precondition for establishing and sustaining local economic development. Partnerships and adjustments of the institutional structures are seen as the main tools in the process of capacity building. Besides the label 'community-led rural development' (Murray and Dunn, 1995), labels like 'community development' (Keane and Ó Cinnéide, 1986) and 'bottom-up partnership approach' (Mannion, 1996) are also used to indicate this approach.

The starting point in the community-led rural development theory is the observation that many rural regions and communities experience genuine difficulties in generating economic development, due to insufficient capacity to solve economic problems, an inappropriate institutional milieu, and lack of political responsibilities. The key for solving these bottlenecks is in the building of a self-help capacity of communities, which refers to organizational expertise of rural communities with regard to group processes, conflict resolution, mediation, leadership, understanding the business of government, and achievement of a shared vision. The ultimate goal is to transform an attitude of apathy and dependency into one of spirit and self-reliance. Or putting it simply: it aims to teach people how to catch a fish rather than presenting them one on a plate (Keane and Ó Cinnéide, 1986:287). Capacity building is a slow process, and often involvement of outside animateurs is required. The process consists of two main elements: the creation of partnerships among actors and the adjustment of the institutional structure. Effective partnerships are those which (Mannion, 1996:5):

- represent and bring together all relevant groups and sectors and enable them to identify and bring forward development possibilities;
- link individual and community development proposals with sources of support and funding;
- consist of private sector representatives who are willing to share power;
- experience and responsibility as equals with community representatives;
- take into account regional or local requirements and initiatives.

Adjustment of the institutional structure is especially needed with respect to the linkages between the local, regional and national authorities, as the community-led development approach requires an institutional structure that encourages and responds to bottom-up initiatives. The initiative for community-led rural development can be with community leaders, but often assistance from outside is necessary, such as partnerships with regional or national authorities, with universities and with developing agencies.

In Bryden's theory (1998) on the potentials of immobile resources for creating competitive advantages in rural areas, it is argued that given the increased mobility of financial and capital flows, skilled labour, information, and other goods and services in the current globalization process, these resources are an unstable basis upon which to build a development strategy for rural areas. Besides, these mobile resources are scarce, implying first, that rural areas have to compete with each other for these resources, and second, that the success of the one can only be achieved at the cost of the other. Hence, Bryden's thesis is that the competitive advantage of rural areas should be based on immobile resources, which are not open for competition.

Immobile resources are resources specific to the locality and which cannot be moved to another location. Some of them are tangible like property, physical infrastructure and natural resources, and others are intangible like knowledge, values and culture. Bryden distinguishes four types of immobile resources:

(1) Social capital: this encompasses the features of social organization such as trust, norms and networks, which can improve the efficiency of society by facilitating co-ordinated actions. Social capital is embedded in relationships among people; it tends to cumulate when it is used and to be depleted when it is not.

(2) Cultural capital: this includes history, traditions, customs, language, music, art and stories, which may be territorially defined as belonging to an area.
(3) Environmental capital: this refers to the actual physical conditions of space of an area. It includes both natural environmental capital (landscape, climate etc.) and built environmental capital (structures of historical significance, physical and tourist infrastructure).
(4) Local knowledge capital: this is about the capacity of the area to generate, sustain and build on formal and informal stocks of knowledge and information.

The immobile resources reveal the opportunities and constraints for local development and also reflect the effectiveness of the local and regional institutional system in handling these opportunities and constraints. Economic development of rural regions can be explained by a combination of tangible and less tangible factors and the way these interact with each other in the local context.

The endogenous development approach supposes the existence of a local growth potential in each region that is waiting to be unlocked. Slee (1994:191) considers endogenous development not so much as 'a concept with clearly defined theoretical roots but more a perspective on rural development, strongly underpinned by value judgements about desirable forms of development'. Slee (1994:193-4) denies the existence of an endogenous development model; he just views it as an exogenous model, in which external forces are the principal determinants of development, but where endogenous forces may colour the nature of the process. In the endogenous model, the emphasis in rural development policies has shifted from a branch-plant strategy towards support for local entrepreneurs, from a single agency activity towards an integrated approach, and from a traditional bureaucratic support structure towards a creation of animators with networking functions, but this does not alter the fundamental nature of the development process. With this criticism we arrive almost as a matter of course at the next approach.

Mix of the exogenous and the endogenous development approach

This approach rejects the polarization of exogenous and endogenous development models and proposes 'an approach of the analysis of rural development that instead stresses the interplay between local and external forces in the control of development processes' (Lowe *et al.*, 1995:87). This approach relates rural development to the process of increasing globalization, due to rapid technological changes in the communications and information sectors. In this changing global context, actors in rural regions are involved in both local networks and external networks, but the size, direction and intensity of networks vary among regions. These networks cover for example links between the various parts within the firm, links among firms, links of firms with local and non-local institutions, and links among institutions. Hence, rural development is considered as a complex mesh of networks in which resources are mobilized and in which the control of the process consists of an interplay between local and external forces (Lowe *et al.,* 1995). Although the concept of innovation is not explicitly mentioned in the mixed exogenous/endogenous approach, it is clear that economic dynamics are derived from the interplay of local and external forces, and so this approach is part of the territorial innovation models.

In adopting the view that rural development is a complex mesh of networks, Lowe *et al.* propose to transform the analysis of economic development of rural regions into an analysis of networks. The focus on networks usefully integrates economic forms with social processes. From the perspective that networks are sets of power relationships and that local and external networks form a geography of networks, the analysis of networks focuses upon questions like (Lowe *et al.,* 1995:100):

*   which actors come to exercise power over others within and through networks;
*   how are local actors drawn into sets of relations and on what terms; what links local actors to external actors;
*   how do external actors effect change and control from a distance?

From the network analysis we can get insights in which particular networks provide beneficial outcomes for rural regions. Some of these networks might be region specific, others might be complex internal/external relations. On the other hand, the network analysis may also provide insights in inequalities and asymmetrics within the networks, which result in a weakening of the position of local actors.

Two simultaneous activities for a development strategy are recommended by Lowe *et al.* (103):

*   try to create linkages between internal networks and institutions, so that 'thick' ensembles arise, which are mutually reinforcing and able to put regions on viable growth trajectories;
*   try to affect the balance of power in local/external networks in such a direction that local actors are enabled to exert control and to retain a reasonable proportion of the value added.

Some regions will not manage to generate development. If regions are trapped in a situation of inequalities and asymmetrics within the networks, a policy goal might be to recast the networks by seeking equity between participants and equality of participation.

Field of force of rural regions

In the RUREMPLO project, we have embraced the mixed exogenous/endogenous approach for the analysis of employment development in rural regions for several reasons:

- although the exogenous approach allows for a local colour of external forces in the development process, it goes beyond endogenous potentials such as regional identity, entrepreneurial climate and attractiveness of the cultural and natural environment;
- in the changing global situation, in which rural regions are involved in various external relationships, the endogenous approach seems to be out of date as these external relationships will affect local development;
- the mixed endogenous/exogenous approach, which sees rural development as a mesh of networks, fits well to the current situation of diverse internal and external relationships of rural regions. The emphasis on the interplay of internal and external forces in the development process offers more perspectives for diverse development trajectories of diversified rural regions than the a priori presuppositions of the exogenous or endogenous approach.

Starting from the mixed exogenous/endogenous approach, the analysis of employment development of rural regions should take the following elements into account:

- identification of the role of the actors in the local networks;
- identification of the role of the actors in the external networks;
- local resources mobilized in the networks;
- external resources transmitted through networks into the rural region.

The analysis of employment development of rural regions can be facilitated by a design of a field of force of rural regions, which includes the elements above (Fig. 2.2). In this design the current global restructuring process, due to rapid technological changes in the communications and information sectors and due to political changes, is taken into account. The changing global situation results in an intensification of the external integration of rural regions. By using the territorial approach, the rural region is presented as a regional economy, which has all kind of exchanges with the external world. Within the rural region, we distinguish three closely related components: local resources, economic activities and actors. The component of local resources refers to physical infrastructure (roads, railways, ports etc.), natural resources (like wood and hydropower) and rural amenities. Economic activities refer to all kind of activities in the agricultural, industrial and service sector. Actors are embodied with capacity (knowledge, skills and attitude) and interact with each other in networks. Moreover, actors can be involved in all kinds of relations with the outside world, like the exchange of products, services and know-how and contacts with policy makers outside the region. Besides, actors are moving into and out of the region. Such migrants generally refer to economic active people, entrepreneurs and retirees. This field of force offers a framework in which factors encouraging and hampering the maintenance or increase of employment opportunities can be revealed.

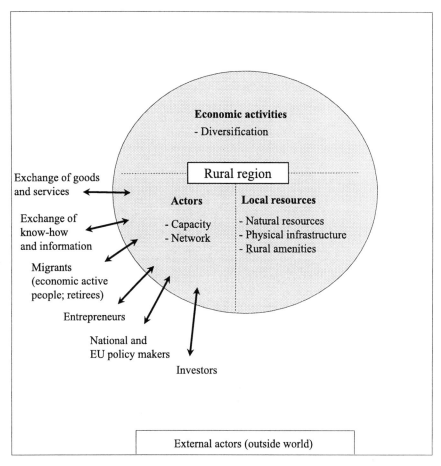

Figure 2.2   Field of force of a rural region
Source: RUREMPLO project.

Local resources
(1) Are local resources (including infrastructure) important for the creation of employment?

Economic activities
(2) In which branches does employment increase (decrease)? What are the properties of these branches?
(3) Does the sectoral mix explain the dynamics in employment growth (stagnation)?
(4) Is employment created in small or large enterprises?
(5) Is employment created in new or existing enterprises?

Actors: analysis of labour supply
(6) Does the education level of the labour force matter in the creation (stagnation) of employment?

Actors: labour market
(7) Is employment hampered by the institutional structure of the labour market?

Actors: analysis of strategies
(8) Does the capacity of actors matter in the creation (stagnation) of employment?
(9) Specify the role of internal and external networks in the creation (stagnation) of employment and give an analysis of which actors come to exercise power over others within and through networks.
(10) Give an identification of the most effective policies and strategies towards maintaining or augmenting employment and indicate their local implementation (indicate why policies and strategies failed in maintaining or augmenting employment).
(11) How do farm households adapt to the situation of decreasing employment in the agricultural sector? What are the perspectives for tourism on the farm, landscape conservation and pluriactivity for farm households?

Figure 2.3   Key issues

Key issues and SWOT analysis

For facilitating the analysis of the field of force, we have made a list with 11 key issues, referring to each of the three main components (Fig. 2.3). Besides, the various forces in the field have been assessed by making a SWOT (Strengths, Weaknesses, Opportunities and Threats) analysis for 1980 (beginning of the period under study) and 1997 (ending year of the period under study). Usually, firms carry out a SWOT analysis in order to assess the prospects of products. However, it has also been used in designing integrated rural development programmes (Moseley, 1996). Insight into the strong and weak points, opportunities and threats of a region provides a better understanding of its problems and perspectives. The items in the SWOT analysis are given in Fig. 2.4.

Strengths and weaknesses
- location of the region (proximity to a large economic centre)
- local resources which favour employment
- education level of the labour force
- low cost labour
- well-developed physical infrastructure
- favourable industry structures
- favourable climate
- favourable amenities
- presence of universities and other major research centres
- specialization of the regional economy
- diversification of the regional economy
- vertical coordination within sectors
- horizontal coordination across sectors
- capacity (knowledge, skills and attitude) of local actors
- capacity of local actors to innovate
- entrepreneurial climate
- internal networks
- external networks
- market 'niches'
- tourism

Opportunities and threats can refer for example to:
- market opportunities (often expressed in relation to a certain sector)
- development of market 'niches'
- development of tourism
- European integration and extension
- political events like GATT/WTO and the transformation process in Central and Eastern European Countries
- improvements in infrastructure (not only in the region itself, but also in other regions such as the construction of a highway, which connects the region with a main economic centre)
- improvement of the access to markets

Figure 2.4   Items of the SWOT analysis

Comparable case studies

Based on the field of force, the key issues and the SWOT analysis, a template for carrying out case studies on employment dynamics in rural regions has been made. The RUREMPLO team has carried out case studies in nine pairs of leading and lagging rural regions in order to analyse employment dynamics in the period 1980-1997. The use of the template for case studies guarantees their mutual comparability. The size of the case study regions reflects more or less that of a functional labour market. The labels leading and lagging have been derived from non-agricultural employment growth in the 1980s and early 1990s (see Chapter 3). We have selected rural regions which were not unique in their development pattern or location, but from which we expected that they could provide insight in the factors behind the process of employment growth/stagnation and lessons for other rural regions. Nevertheless, the selected regions reflect a wide range of characteristics with regard to their location, industrial tradition and physical structure. The case studies have been based on analyses of statistics, literature and interviews with key actors. The main findings of the case studies are reported in Chapters 4-12.

Concluding remarks

In this chapter the focus has been on finding a suitable approach for the analysis of employment dynamics in rural Europe. In the discussion of the various views on rurality, we have concluded that the territorial approach to rurality is the most useful for our purpose. This enables us to consider rural space as a territorial entity which covers a regional economy with agricultural, industrial and services activities. By doing so, we can take all relevant economic relationships into account. Our choice implies that we distinguish a large number of rural regional economies in Europe, which we have briefly labelled as 'rural regions'. Usually, the size of such regions reflects that of a labour market area.

In the review of regional economic growth theories, we have made a distinction into four groups depending on the factors in the production function: traditional models, pure agglomeration models, local milieu models and territorial innovation models. We have shown that the different stages in the debate on economic development in rural regions – the exogenous approach, the endogenous approach and the mixed exogenous/endogenous approach – have a close relationship with respectively pure agglomeration models, local milieu models and territorial innovation models.

The mixed exogenous/endogenous approach has been embraced as a useful approach for analysing employment dynamics in rural regions in Europe. This approach sees rural development as a mesh of networks of local and external actors, in which resources are mobilized and in which the control of the process consists of an interplay between local and external forces. Based on this approach, we have designed a conceptual framework for analysing employment dynamics: the field of force of a rural region. Within the rural region, we have distinguished three closely related components: local resources, economic activities and actors. The component of local resources refers to physical infrastructure (roads, railways, ports etc.), natural resources (like wood and hydropower) and rural amenities. Economic activities refer to all kind of activities in the agricultural, industrial and service sector. Actors are embodied with capacity (knowledge, skills and attitude) and interact with each other in networks. Moreover, actors can be involved in all kinds of relations with the outside world like the exchange of products, services and know-how, contacts with policy makers outside the region, and migration. This field of force offers a framework in which factors encouraging and hampering the maintenance or increase of employment opportunities can be revealed.

References

Becattini, G. (1987) Il distretto industriale marshalliano: cronaca di un ritrovamento. *Mercato e forze locali*, Il Mulino.

Blanc, M. (1997) Rurality: concepts and approaches. Paper presented at the 48th Seminar *'Rural restructuring within developed economies'* of the European Association of Agricultural Economists, Dijon, 20-21 March.

Borgstein, M.H., Graaff, R.P.M. de, Hillebrand, J.H.A., Scherpenzeel, J.F., Sijtsma F.J. and Strijker D. (1997) *Ketens en plattelandsontwikkelingen (Chains and rural development.* NRLO-rapport nr. 97/34, The Hague.

Bryden, J.M. (1998) Development strategies for remote rural regions: what do we know so far? Paper presented at the *OECD international conference on remote rural areas – developing through natural and cultural assets,* Albarracin, Spain, 5-6 November.

Camagni, R.P. (1992) *Final report of the research project on development prospects of the Community's lagging regions and the socio-economic consequences of the completion of the internal market; An approach in terms of local 'milieux' and innovation networks.* GREMI, Milan.

Camagni, R. (1995a) Global network and local milieu: towards a theory of economic space. In: Conti, S., Malecki, E.J. and Oinas, P. (eds) *The industrial enterprise and its environment: spatial perspectives.* Avebury, Aldershot, pp. 195-214.

Camagni, R. (1995b) The concept of innovative milieu and its relevance for public policies in European lagging regions. *Papers of the Regional Science Association* 74-4, 317-340.

Capello, R. (1996) Industrial enterprises and economic space: the network paradigm. *European Planning Studies* 4-4, 485-498.

Errington, A. (1994) The peri-urban fringe: Europe's forgotten rural areas. *Journal of Rural Studies* 10-4, 367-375.

Halfacree, K.H. (1993) Locality and social representation: Space, discourse and alternative definitions of the rural. *Journal of Rural Studies* 9-1, 23-37.

Healy, M.J. and Ilbery, B.W. (1990) *Location and change; Perspectives on economic geography.* Oxford University Press, Oxford.

Hoggart, K., Buller, H. and Black, R. (1995) *Rural Europe; Identity and change.* Arnold, London.

Iacoponi, L., Brunori, G. and Rovai, M. (1995) Endogenous development and the agroindustrial district. In: Ploeg, J.D. van der and Dijk, G. van (eds) *Beyond modernisation; The impact of endogenous rural development.* Van Gorcum, Assen, pp. 28-69.

Keane, M.J. and Ó Cinnéide, M.S. (1986) Promoting economic development amongst rural communities. *Journal of Rural Studies* 2-4, 281-289.

Keating, M. (1998) *The new regionalism in Western Europe; Territorial restructuring and policitcal change.* Edward Elgar, Cheltenham and Northampton.

Lambooy, J.G., Wever, E. and Atzema, O.A.L.C. (1997) *Ruimtelijke economische dynamiek; Een inleiding in de theoretische aspecten van de economische geografie (Spatial economic dynamics; An introduction into the theoretical aspects of economic geography).* Coutinho, Bussum.

Lowe, P., Murdoch, J. and Ward, N. (1995) Networks in rural development beyond exogenous and endogenous models. In: Ploeg, J.D. van der and Dijk, G. van (eds) *Beyond modernisation; The impact of endogenous rural development.* Van Gorcum, Assen, pp. 87-106.

Malecki, E.J. (1991) *Technology and economic development: the dynamics of local, regional, and national change.* Longman, Essex.

Mannion, J. (1996) Strategies for local development in rural areas: the 'bottom-up' approach. Paper prepared for the *European Commission Conference on Rural Development,* Cork, 7-9 November.

Molle, W. and Cappellin, R. (1988) *Regional impact of community policies in Europe.* Avebury, Aldershot.

Moseley, M.J. (1996) Baseline studies for local rural development programmes: towards a methodology. *Planning Practice and Research* 11, 19-36.

Murray, M. and Dunn, L. (1995) Capacity building for rural development in the United States. *Journal of Rural Studies* 11-1, 89-97.

Picchy, A. (1994) The relations between central and local powers as context for endogenous development. In: Ploeg, J.D. van der and Long, A. (eds) *Born from within; Practice and perspectives of endogenous rural development.* Van Gorcum, Assen, pp. 195-203.

Porter, M.E. (1990) *The competitive advantage of nations.* Macmillan Press, London.

Pratt, A.C. (1996) Discourses of rurality: loose talk or social struggle? *Journal of Rural Studies* 12-1, 69-78.

Rijswick, C.W.J. van (1997) *Regional economic theory; Its usefulness for explaining economic development in rural EU regions.* LEI-DLO, MA Thesis (internal document), The Hague.

Saraceno, E. (1994) Alternative readings of spatial differentiation: The rural versus the local economy approach in Italy. *European Review of Agricultural Economics* 21-3/4, 451-474.

Slee, B. (1994) Theoretical aspects of the study of endogenous development. In: Ploeg, J.D. van der and Long, A. (eds) *Born from within; Practice and perspectives of endogenous rural development.* Van Gorcum, Assen, pp. 184-194.

Storper, M. (1995) The resurgence of regional economies, ten years later: the region as a nexus of untraded interdependencies. *European Urban and Rural Studies* 2-3, 191-221.

Statistical Analysis of Employment Growth in Rural Regions of the EU, 1980–1995

3

Roberto Esposti, Frans E. Godeschalk, Tuomas Kuhmonen, Jaap H. Post, Franco Sotte and Ida J. Terluin

Introduction

Employment growth varies among EU regions. From a recent analysis, it appears that there were dynamic rural regions which showed an employment performance above the national average during the 1980s, and that there were also rural regions whose employment growth lagged behind (OECD, 1996). This observation directly results in the question: why do some rural regions perform better than others do? Can the sectoral mix of employment explain these differences? Or are these differences the result of factors like local resources, natural and cultural amenities, entrepreneurial tradition, work ethics, public or private networks, a set, which can be referred to as 'territorial dynamics'?

In this chapter we examine whether rural regions in the EU with a relatively high (low) employment growth in the 1980s and early 1990s have some common socio-economic characteristics, which can contribute to the explanation of their employment performance. As common socio-economic characteristics we use only those that can easily be quantified. We have grouped these variables around the headings 'economic activities', 'local resources' and 'supply of labour'. These headings are derived from the field of force of a rural region, as explained in the previous chapter. In order to make a further delimitation of the research, we have listed a number of hypotheses, which will be investigated. These are as follows:

- Rural regions with employment growth tend to have population growth as well.
- Rural regions with a higher share of employment in agriculture tend to show a lower non-agricultural employment growth.
- Employment growth tends to be higher in rural regions with a higher GVA/worker.
- Rural regions with a higher endowment of infrastructure tend to show a higher employment growth.
- In rural regions the education level of the population is lower than in urban regions.

- In rural regions with a higher education level of the population, the increase in employment is higher than in other rural regions.
- In rural regions with a higher education level of the population, the economy is more diversified than in other rural regions.
- Rural regions with employment growth tend to have higher participation rates than rural regions in which employment growth stagnates.
- Rural regions with high unemployment rates tend to have lower participation rates than rural regions with low unemployment rates.
- In rural regions with a higher education level of the population, participation rates tend to be higher.
- The percentage of farmers with other gainful activities in rural regions with employment growth tends to be higher than in rural regions in which employment growth stagnates.

We approach these hypotheses by making correlations between pairs of variables. We only search for a relationship between variables, without explaining deeper reasons behind. Such a further examination goes beyond the scope of this statistical analysis. Nevertheless, several hypotheses, which are supported by the analysis, will be further investigated in the next chapters of this book.

Plan of this chapter

We base our analysis on 465 regions in the EU15, which we have divided into three groups: most rural regions, intermediate regions and most urban regions. Within each of these three groups, we have made a further distinction into leading and lagging regions. This clustering of regions is discussed in the next section. In order to carry out this statistical analysis of employment development, we have collected a large amount of data at regional level. This database is also elaborated in the next section. As a next step, successively attention is paid to the employment pattern in the various clusters of regions, the sectoral structure and GDP, to the extent to which transportation infrastructure matters in employment growth, to differences in the structure of firms and farms, and to various aspects of the supply of labour and their relationship with employment dynamics. In the last section some concluding remarks are made.

Clustering of regions and used data

In the analysis of employment development in rural regions of the EU, we base the clustering on a regional level which reflects more or less the size of a functional labour market. For Belgium, Germany, Luxembourg and the Netherlands, this implies that we work at NUTS2 level; in the other EU Member States the NUTS3 level approaches the size of a functional labour market. Due to lack of data at NUTS3 level, we used also NUTS2 level for Greece and Portugal. For Austria the NUTS classification was inconvenient for analytical purposes, owing to specific commuting patterns in mountainous areas. So for this country we delineated 32 regions, which are more or less homogeneous in natural conditions, economic development and political background. Also for Finland because of lack of data at NUTS3 level, a classification into 12 regions has been designed for the purpose of this study. The total number of regions in the study is 465. The use of NUTS2 and NUTS3 regions implies that the size of regions varies, which may affect the results.

Degree of rurality

In this study three types of regions are distinguished:

- most rural regions;
- intermediate regions;
- most urban regions.

These three types are based on population density at local and regional level, derived from the methodology used in the OECD classification of predominantly rural, significantly rural and predominantly urban regions (OECD, 1994). The territorial scheme distinguishes two hierarchical levels: the local community level and the regional level. Local communities are basic administrative units with a very detailed grid like *cantons* in France, districts in the UK and *comuni* in Italy. Regions are larger administrative units or functional zones with a less detailed grid like *aemter* in Denmark, *provincias* in Spain and *provincies* in Belgium and the Netherlands. If population density in local communities is less than 150 inhabitants per square kilometre, the community is classified as 'rural'; if population exceeds 150 inhabitants per square kilometre as 'urban'. As a second step, regions are divided into three groups. If more than 50% of the population of the region lives in rural local communities, the region is classified as 'most rural'; if between 15 and 50% of the population of the region lives in rural local communities, the region is classified as 'intermediate'; and if less than 15% of the population of the region lives in rural local communities, the region is classified as 'most urban'. Moreover, if regions include a city of 200,000 inhabitants or more, the region is classified as intermediate; if regions include a city of 500,000 inhabitants or more, the region is classified as most urban region. The distribution of regions over the three classes of rurality is given in Table 3.1 and graphically presented in Fig. 3.1.

Leading and lagging regions

Within the groups of most rural and intermediate regions, we made a further distinction into leading, average and lagging regions based on the performance of non-agricultural employment growth in the 1980s and early 1990s. A region is considered to be leading if the growth rate of non-agricultural employment was 0.5 percentage points above the national growth rate; on the other hand, a region is considered to be lagging if the growth rate of non-agricultural employment was 0.25 percentage points below the national growth rate. By relating the regional growth rate to the national one, regional growth rates are corrected for differences

Table 3.1 Classification of EU15 regions to the degree of rurality

Degree of rurality	Number of regions	Percentage
Most rural regions	195	42
of which:		
leading	58	12
lagging	49	11
Intermediate regions	164	35
of which:		
leading	57	12
lagging	36	8
Most urban regions	106	23
of which:		
leading	23	5
lagging	40	9
Total	465	100

among the absolute level of national growth rates. It has to be emphasized that the labels leading and lagging are only derived from employment performance, and that leading regions may be less successful with regard to other indicators like GDP per capita, GDP growth and unemployment rates. Moreover, it appears that the growth rate of employment can change considerably when using another period. This implies that if a region is labelled as lagging, this is not necessarily a permanent situation, but that it can change.

Looking at the map of leading and lagging most rural and intermediate regions (Fig. 3.2), it can be seen that both leading and lagging regions are often located in groups. Leading regions tend to be located at coastlines, whereas lagging regions tend to be located in the inner parts of countries like Spain, France and Austria; both in North and South Italy; in the western part of Greece; and in the Scandinavian countries. The location of the leading regions indicates that access to waterways/ports and coastlines, which are attractive for tourists, can have a positive impact on employment performance. However, there are also many regions with a coastal location which are not leading.

Data

The main data source used was Eurostat REGIO. Agricultural variables have been derived from Eurostat Farm Structure Survey (FSS). Data concerning tourism, education level, number of enterprises and concentration of population have been collected by the participants in the RUREMPLO project. The data on the degree of rurality have been derived from the OECD Rural Data Survey but have been modified for the purpose of our study. Data for the three new EU Member States

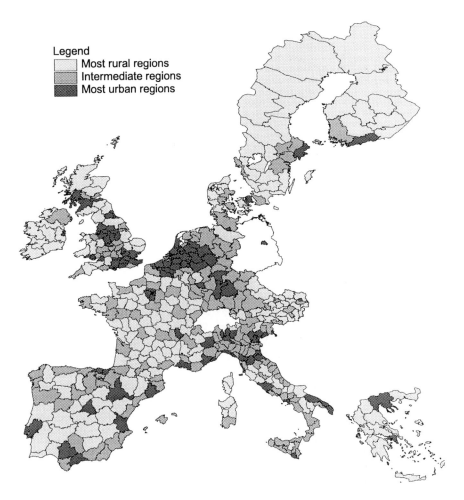

Figure 3.1 Degree of rurality of the EU15 regions
Source: RUREMPLO project.

(Finland, Sweden and Austria) have also been provided by the participants. With the exception of the agricultural variables, data should cover the years 1980, 1985, 1990 and a recent year. The agricultural variables are given for 1989/1990. The advantage of using Eurostat data sources is that data are harmonized. However, it appeared that within the Eurostat data there were several gaps, in particular for the NUTS3 regions. So repeatedly we had to switch to national sources. This implies that different definitions for data have been used for a number of variables. Sometimes no data at all were available (see Esposti *et al.* (1999) for an extensive description of the data). The lack of data and the use of different definitions limit the results of our analysis to some extent.

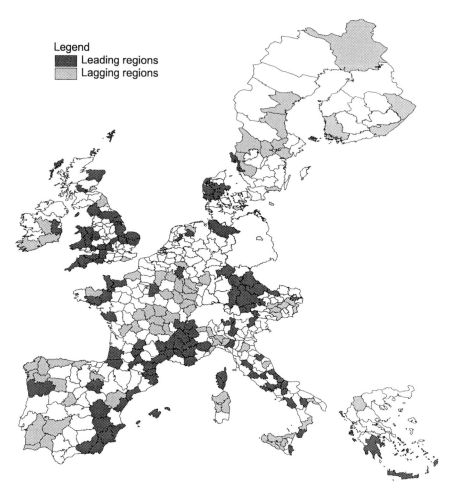

Figure 3.2   Leading and lagging most rural and intermediate regions in the EU
Source: RUREMPLO project.

Employment pattern

In all groups of regions, employment in agriculture decreased and employment in services increased during the 1980s (Table 3.2). This development reflects the common pattern of employment development in advanced countries. Besides, the leading most rural and intermediate regions showed an increase in industrial employment, which is quite surprising as it is usually taken for granted that industrial employment declines in EU countries. A second striking point in regional employment growth is that employment in the leading most rural and intermediate regions increased at a higher rate than that in the most urban regions. This was the result of two developments: the absence of a decrease in

Table 3.2 The annual rate of change in employment and population by types of regions in about '1980-93' (%)

Regions	Employment growth					Population growth
	Total	Agricul-ture	Indus-try	Servi-ces	Non-agri-culture	
Leading most rural	0.8	-3.7	0.2	2.1	1.4	0.51
Leading intermediate	1.0	-3.6	0.3	2.4	1.6	1.09
Lagging most rural	-0.7	-4.6	-1.8	1.3	0.1	-0.06
Lagging intermediate	-0.7	-4.8	-2.0	1.0	-0.2	0.06
Most rural regions	0.0	-4.1	-0.8	1.7	0.8	0.26
Intermediate regions	0.4	-3.9	-0.6	1.7	0.8	0.52
Most urban regions	0.5	-3.3	-1.1	1.5	0.6	0.32
All regions[a]	0.4	-3.9	-0.9	1.6	0.7	0.37

a) Average of regions above.

employment in industries in the most urban regions on the one hand, and a higher rise in employment in services in the most rural and intermediate regions on the other hand.

Employment dynamics related to population growth

The leading most rural and intermediate regions showed also a population growth in the 1980 and early 1990s, whereas population growth in the lagging ones stagnated (Table 3.2), indicating that employment growth is accompanied by population growth. The population in the leading intermediate regions increased at a higher rate than that in the leading most rural regions, but population growth in both groups of leading regions was above that in urban regions.

Convergence sectoral structure of employment

Differences in the sectoral structure of employment between rural and urban regions decreased during the study period (Table 3.3). In the 1990s, the share of employment in agriculture varies from 13% in lagging most rural regions to 2.5% in most urban regions; the share of employment in industries is almost the same in all groups (about 30%); whereas the share of employment in services ranges from 58% in lagging most rural regions to over 67% in most urban regions.

Table 3.3 The sectoral structure of employment by types of regions (%)[a]

Regions	Agriculture		Industries		Services	
	1980s	1990s	1980s	1990s	1980s	1990s
Leading most rural	15.3	9.7	31.1	29.4	53.6	60.9
Leading intermediate	13.2	7.7	35.9	32.8	51.0	59.4
Lagging most rural	20.9	13.0	33.0	28.9	46.0	57.8
Lagging intermediate	13.4	7.8	36.7	30.9	49.9	61.3
Most rural regions	20.0	12.5	32.1	29.4	47.9	57.9
Intermediate regions	11.5	6.8	37.3	33.2	51.2	59.9
Most urban regions	3.7	2.5	35.7	30.1	60.6	67.4
All regions	9.2	5.7	35.6	31.0	55.2	63.3

a) Figures refer to the beginning of the study period (1980s) and to the end of the of the study period (1990s).

Table 3.4 Correlation coefficients between share of agriculture at the beginning of the 1980s (%) and employment growth (% p.a. in '1980-1993')

	Total employment	Non-agricultural employment	Agricultural employment
Leading most rural	-0.68	-0.08	-0.18
Leading intermediate	-0.46	0.04	-0.03
Lagging most rural	-0.77	0.04	-0.26
Lagging intermediate	-0.52	-0.05	-0.08
Most rural regions	-0.58	-0.06	-0.25
Intermediate regions	-0.28	0.04	-0.07
Most urban regions	-0.06	0.24	-0.14
All regions	-0.42	0.08	-0.18

Higher share agriculture implies lower growth total employment

Rural regions, with a higher share of employment in agriculture in the beginning of the period, tend to have a lower growth of total employment in the 1980s since a higher share of employment in agriculture induces a higher decline in agricultural employment (Table 3.4). This correlation is stronger for most rural regions relative to intermediate regions. There is no relationship between the share of employment in agriculture and the growth rate of non-agricultural employment.

Table 3.5 GDP per capita in the early 1980s and early 1990s (in ECU's)[a]

Region	GDP/capita		Index (all regions = 100)	
	1980s	1990s	1980s	1990s
Leading most rural	7,000	13,700	89	88
Leading intermediate	7,100	14,600	90	94
Lagging most rural	5,900	11,900	75	77
Lagging intermediate	7,600	14,400	96	93
Most rural regions	6,200	12,500	79	80
Intermediate regions	7,600	15,000	96	97
Most urban regions	8,700	17,000	110	109
All regions	7,900	15,500	100	100

a) The Finnish, Irish, Italian and Swedish regions are excluded, together with two French regions. The period covered is 1980-1993 for Belgium and Denmark, 1980-1990 for Germany, 1980-1993 for Greece, Spain, France (1986-1993 for six regions), Luxembourg and the Netherlands (1986-1993 for three regions), 1988-1992 for Austria, 1980-1993 for Portugal and 1981-1993 for the UK.

GDP per capita in intermediate regions close to the EU average

GDP per capita in most rural regions is about 20% below the EU average, GDP per capita in intermediate regions is less than 5% below, and that in most urban regions about 10% above the EU average (Table 3.5). These deviations from the EU average are rather stable in the 1980s. GDP per capita in lagging most rural regions is below that in the leading most rural regions, although the gap narrows a little in the 1980s. On the contrary, in 1980 GDP per capita in lagging intermediate regions was above that in leading regions. Since GDP per capita in leading intermediate regions increased at a higher rate, in the 1990s GDP per capita in these regions exceeded that in the lagging ones. In the interpretation of the development of GDP per capita among the different rurality groups, a country bias has to be taken into account: the groups are composed of both regions with a relatively high GDP per capita and regions with a relatively low GDP per capita.

Labour productivity affects employment growth in most rural regions

In the group of most rural regions, there is a strong positive relationship between the level of labour productivity (measured as gross value added (GVA) per worker) at the beginning of the 1980s and total employment growth in the period '1980-1993' (Table 3.6). However, in the group of intermediate and most urban regions this relationship is negligible. This could be explained by the relatively large agricultural and industry sectors in most rural regions, in which global markets and competitiveness play an important role.

Table 3.6 Correlation coefficients between GVA/worker at the beginning of the 1980s and employment growth in (% p.a. in '1980-1993')[a]

	Total employment	Non-agricultural employment	Agricultural employment
Most rural regions	0.67	0.35	0.48
Intermediate regions	0.04	0.16	0.08
Most urban regions	-0.14	-0.43	0.16
All regions	0.40	0.04	0.28

a) Calculation is based on relative figures: GVA/worker and employment growth are related to the national average. The French, German, Irish, Italian, Portuguese, Swedish and UK regions are excluded.

Infrastructure

Infrastructure endowment is usually considered as one of the most important local resources allowing people and goods to move out and into the region. There are many different types of infrastructures, which can be strategically relevant for the development of a region; however, due to the difficulty in finding data at regional level, we concentrate only on three categories of transportation infrastructure: motorways, other roads and railroads. We calculated a very simple endowment index (*EI*) given by:

$$EI_i = \frac{L_i}{TA_i}$$

where L_i is the length of motorways, other roads and railroads respectively, and TA_i the total area of the *i*-th region.

On average the infrastructure endowment of most rural regions is lower than intermediate and most urban regions for each of the three transportation categories, while urban regions are more endowed than intermediate regions (Table 3.7). The endowment of the urban regions is in each case more than three times the average of rural regions. Besides, leading most rural and intermediate regions tend to have a higher infrastructure endowment than the lagging ones, with the exception of railway endowment in most rural regions.

Firms and farms

The size structure of firms in rural and urban regions is more or less the same: about 92% of all firms are small (fewer than 10 employees), about 7% of the firms have 10-49 employees, and just over 1% of firms have more than 50 employees (Table 3.8). This distribution was stable during the 1980s.

Table 3.7 Density of transportation infrastructure (EI_i), 1990[a)]

	Motorways	Other roads	Railroads
Leading most rural	0.02	0.95	0.05
Leading intermediate	0.04	1.28	0.02
Lagging most rural	0.01	0.33	0.03
Lagging intermediate	0.02	1.10	0.09
Most rural regions	0.01	0.35	0.03
Intermediate regions	0.02	1.00	0.04
Most urban regions	0.05	2.06	0.10
EU average[b)]	0.04	0.87	0.05

a) Analysis of motorways and other roads is based on data for Belgium, Greece, Spain, Luxembourg, Netherlands, Finland and Sweden and some regions from Austria and Portugal; analysis of railways is based on data for Belgium, Spain and Luxembourg and some regions from other countries; b) Average of all EU member states.

Table 3.8 The size structure of enterprises by types of the regions (%)[a)]

Region	<9 employees		10-49 employees		>50 employees	
	1980s	1990s	1980s	1990s	1980s	1990s
Most rural regions	91.5	91.4	7.1	7.3	1.4	1.3
Intermediate regions	93.1	93.3	5.9	5.8	1.0	0.9
Most urban regions	91.4	91.9	7.2	6.8	1.5	1.3
All regions	92.0	92.3	6.7	6.5	1.3	1.2

a) The Danish, German, Spanish, French, Irish, Luxembourg, Portuguese and the UK regions are excluded.

80% of EU farms in most rural and intermediate regions

From the total of 8.8 million farms in the EU in 1990, about 40% were located in most rural regions and 40% in intermediate regions (Table 3.9). The average farm size in lagging most rural regions is about 17 ha and that in leading most rural regions about 12 ha. The farm size in leading and lagging intermediate regions is almost the same and amounts to 7.5 ha. In lagging most rural regions about two-thirds of the utilized agricultural areas is classified as Less Favoured Area (LFA), in leading most rural regions the share of LFA is 60%, and in intermediate regions it amounts to 50% (Esposti *et al.*, 1999:32).

Table 3.9 The number of farms, forest land and utilized agricultural area (UAA), 1989/1990

Region	Farms, 1,000	Farms per capita ha	Forest, ha per capita[a]	UAA, ha per capita[b]	UAA, ha per farm
Leading most rural	935	0.04	0.6	0.9	12.4
Leading intermediate	1,430	0.03	0.2	0.3	7.6
Lagging most rural	740	0.05	1.7	1.1	17.4
Lagging intermediate	800	0.04	0.3	0.3	7.3
Most rural regions	3,520	0.05	1.3	1.0	13.7
Intermediate regions	3,510	0.03	0.2	0.4	9.9
Most urban regions	1,760	0.01	0.1	0.1	8.2
All regions	8,790	0.03	0.4	0.3	11.0

a) Except for the Danish, Irish and the UK regions; b) Except for the Danish, French, Portuguese and the UK regions.

The number of farms per capita describes the importance of farms as a place to live (biased by the varying size of families). This number is clearly the highest in most rural regions (about 50 per 1000 inhabitants) and logically the lowest in the most urban regions (about 10 per 1000 inhabitants) (Table 3.9). The number of farms per capita is higher in the lagging regions than in the leading regions. The most rural regions are also characterized by the highest endowment of forest land and agricultural land per capita. The lagging most rural regions have much higher endowment rates than the leading most rural regions, whereas in the intermediate regions the difference is rather small.

Labour supply

In this section we will deal with some main characteristics of labour supply like education level, participation rates, unemployment and part-time employment, trying to identify differences between different rurality degrees and between leading and lagging regions. These characteristics and their dynamics are also related to the regional employment performance.

Lower education level in rural regions

The education level in rural regions is below that in urban regions. In the early 1990s, 6% of the population in the group of most rural regions had a tertiary level education, against 8% in the group of intermediate regions and 10% in the group of most urban regions (Table 3.10). The share of the upper secondary level is almost the same in the three rurality classes, whereas the group of the most rural regions has the highest share of population with a lower educational attainment.

Table 3.10 Share of population by education level, 1991 (%) and various correlations with the education level[a]

	Lower secondary	Upper secondary	Tertiary	Correlation coefficients Education and empl. growth[b]	Education and diversification[c]
Leading most rural	58	36	6	-0.16	0.84
Leading intermediate	49	43	8	0.27	0.59
Lagging most rural	53	41	6	0.42	0.62
Lagging intermediate	55	37	8	-0.26	0.69
Most rural regions	55	39	6	-0.13	0.85
Intermediate regions	53	39	8	0.21	0.59
Most urban regions	52	38	10	-0.05	0.68
EU average	53	39	8	0.01	0.80

a) Derived by using the OECD classification: lower secondary level (ISCED 1-2); upper secondary (ISCED 3); tertiary level (ISCED 5-7). Table is based on data for Italy, Netherlands, Austria, Finland and Sweden and some regions of Greece, Spain and France; correlation in last column covers 31% of the regions; b) Correlation is calculated for the (relative) rate of non-agricultural employment growth and the share of secondary and tertiary level education; c) Correlation is calculated for the share of population with upper secondary or tertiary level education and share of industry and services in GVA, 1991/1993.

The evidence that the higher educated labour force tends to be concentrated in urban regions is interesting in the analysis of labour markets, but has no single explanation. It can be explained by the presence of better working opportunities for highly educated labour force in urban regions. Next, young people who want a higher education usually have to attend schools and universities in urban regions, and they frequently tend to stay there after finishing their studies because they cannot find a job or due to lack of jobs in rural regions for people with a higher education. A third reason is that the share of agriculture in employment in rural regions is higher than in urban regions. Agricultural workers have often a relatively low education level.

Employment performance and education level: no relationship

There is no difference in the share of population with a third level education between leading and lagging most rural and intermediate regions (Table 3.10). Leading intermediate regions have a higher share of population with upper secondary education level than the lagging ones. However, the opposite applies for the most rural regions. In order to asses the relationship between employment dynamics and education level of population, we have carried out an OLSQ regression analysis of the relative non-agricultural employment growth with the share of secondary and tertiary level education. Considering all the available regions, no correlation emerges (Table 3.10). However, if we look to the subgroups of regions, we can notice some interesting, although controversial results. The correlation is positive in intermediate regions, while it is negative in most rural and in most urban regions (although the correlation is very weak). In any case, we have to conclude that there is not enough empirical evidence to state that there is some regular relationship between employment performance and education level of population in rural regions.

Higher education level related to a more diversified economy

The relationship between the education level of the population and the diversity of the economy has been analysed by using an OLSQ regression analysis. The diversity of the economy is considered here as the share of industries and services in GVA. It appears that in most rural and intermediate regions with a higher diversification of the economy, the share of population with an upper secondary and tertiary level education is also higher (Table 3.10). However, it has to be noted that the analysis does not reveal the sequence of developments in the regions: does a higher educated population result in diversity or does a diversified economy attract higher educated workers?

Small differences between activity rates in rural and urban regions

The activity rate can be defined as the share of the working age population (i.e., 15-64 years) in the total population and depends on the age structure of the population. Activity rates slowly increase going from most rural regions to the urban ones, but differences are rather small (Table 3.11). This can be explained by a relatively higher share of young people in urban regions and a relatively higher share of older people in rural regions.

Participation rate does not increase with higher employment growth

The participation rate – defined as the ratio of the working population (employed and unemployed persons) and the working age population – in the group of most rural regions exceeds that in the group of intermediate and most urban regions (Table 3.11). The difference can be explained by different socio-economic behaviour, in particular of women and of young people; actually, students and non-working women are the most relevant components that lower the participation rate. Another important determining factor for the level of the participation is the expectation of finding a satisfying job: the difficulty in finding it can have a discouraging effect. Hence it could be expected that in regions with growing numbers of jobs the participation rate increases. However, the correlation between the regional participation rate and relative non-agricultural employment growth is very weak (Table 3.11). Therefore, there is no clear evidence of a linkage between participation rates and recent employment

Table 3.11 Activity rate, participation rate and various correlations with the participation rate, 1990[a]

	Activity rate	Participa-tion rate	Correlation coefficients		
			Part. rate and non-agr. empl. growth[b]	Part. rate and unempl. rate[c]	Part. rate and education level[d]
Leading most rural	0.66	0.66	0.01	-0.61	0.39
Leading intermediate	0.67	0.65	-0.07	-0.70	0.15
Lagging most rural	0.64	0.68	0.31	-0.76	0.50
Lagging intermediate	0.67	0.63	-0.09	-0.32	0.31
Most rural regions	0.65	0.67	0.03	-0.68	0.24
Intermediate regions	0.67	0.64	0.06	-0.58	0.23
Most urban regions	0.68	0.64	-0.09	-0.41	0.33
EU average	0.66	0.67	0.01	-0.58	0.30

a) Analysis is based on all countries, except for Denmark, France and the UK; b) Correlation based on the regional participation rate (1990) and (relative) non-agricultural employment growth; c) Correlation based on the regional participation rate (1990) and unemployment rate in 1990; d) Correlation based on the regional participation rate (1990) and the share of population with upper secondary and third level education, calculated for regions of Italy, Netherlands, Austria, Finland and Sweden and some regions of Greece and Spain.

dynamics. If we consider on the other hand the correlation between participation rate and unemployment rate, the coefficient is significantly negative (Table 3.11); furthermore, it is maximum in rural regions, while it is weaker in most urban regions. Interestingly, the negative correlation is particularly significant in lagging most rural regions. So from this point of view the discouraging effect of unemployment on the participation rate seems to be confirmed and turns out to be particularly relevant in rural regions.

If we consider the relationship between the education level and participation rate, correlation calculations show that a higher education level has an upward effect on the participation rate. There is no relevant difference between the various groups of regions in this correlation (Table 3.11).

Lagging regions tend to have higher unemployment rates

National unemployment rates show for all EU countries a cyclical pattern in the period 1985-1995: they are high in 1985, reach a low in about 1990, and then rise again, Austria and Greece being the exceptions (Esposti *et al.*, 1999: 67-68). This cyclical pattern can also be perceived for the three rurality groups (Table 3.12).

Table 3.12 Unemployment rates, 1985-1995[a]

	1985	1990	1995	Unemployment rate (1995) in regions with a:	
				High share agriculture[b]	Low share agriculture[b]
Leading most rural	0.10	0.07	0.09	0.11	0.09
Leading intermediate	0.12	0.09	0.11	0.14	0.11
Lagging most rural	0.11	0.09	0.13	0.14	0.07
Lagging intermediate	0.11	0.10	0.14	0.14	0.10
Most rural regions	0.10	0.08	0.11	0.13	0.08
Intermediate regions	0.11	0.09	0.11	0.14	0.10
Most urban regions	0.11	0.08	0.10	0.14	0.09
All regions	0.10	0.08	0.11	0.14	0.09

a) Data cover all regions, except some Austrian and Portuguese regions; b) High share: share of agriculture in total employment >10%; low share: share of agriculture in total employment <10%; for most urban regions >5% and <5% respectively.

Unemployment rates hardly differ among most rural, intermediate and most urban regions. However, the unemployment rate in lagging most rural and intermediate regions is higher than in the leading regions in the 1990s. Moreover, the unemployment rate tends also to be higher in regions with a higher share of employment in agriculture. This fact confirms the idea that in many cases high employment rates in agriculture are an indicator of hidden unemployment: if there are few job opportunities in other sectors, agriculture continues to play a role in employing potentially unemployed labour force. Considering the cyclical behaviour of the unemployment rate in the period 1985-1995, it can be seen that leading most rural and intermediate regions had a higher unemployment rate in the beginning of the period, while for the lagging regions the highest rate was found at the end of the period. This implies that leading rural regions were able to reduce unemployment rates, whereas lagging regions failed. However, this pattern may vary among countries as rather large differences in the level of unemployment rates exist.

Hardly any difference in share of women in total employment among regions

The evolution of the female labour force also contributes to the development of the working population. The share of women in total employment is slightly over one-third: the share varies from 34% in the group of most rural regions to 37% in the group of the most urban regions (Table 3.13). However, among countries these differences are considerable, varying from 29% in Luxembourg to 48% in

Table 3.13 Share of female employed in total employment, 1991[a]

	Total	Agriculture	Industry	Services
Leading most rural	0.37	0.41	0.20	0.45
Leading intermediate	0.37	0.39	0.23	0.44
Lagging most rural	0.37	0.29	0.20	0.49
Lagging intermediate	0.36	0.33	0.18	0.44
Most rural regions	0.35	0.33	0.20	0.47
Intermediate regions	0.36	0.36	0.21	0.44
Most urban regions	0.37	0.33	0.22	0.45
EU average	0.41	0.34	0.24	0.50

a) Analysis is based on data for Belgium, Germany, Spain, Ireland, Italy, Luxembourg, Netherlands and Finland and on some regions from Greece, Portugal and Sweden.

Finland. These differences show that female participation is affected by country specific characteristics. Among sectors there is some difference: the share of women in total employment in agriculture is about 34%, in industries 21% and in services 45%. The share of women in total employment in leading and lagging rural regions is about the same, although there are some small differences in the sectoral mix of female participation.

Few part-time jobs in most rural regions

Women can be discouraged from participating if there are not enough opportunities to find an appropriate job. The discouraging effect can be more relevant when the labour market does not offer job opportunities such as part-time jobs that allow women to continue to carry out their traditional care activities. In most rural regions there are fewer part-time jobs than in intermediate and urban regions: the share of part-time jobs in total employment rises from 8% to 17% (Table 3.14). Here again large variations exist among countries: shares vary from 5% in Spain, Italy and Greece to 36% in the Netherlands. The agricultural sector offers most part-time jobs in the most rural regions, while the service sector provides most part-time jobs in the intermediate and most urban regions. It is striking that lagging most rural regions have few part-time jobs, which is mainly caused by a relatively small number of part-time jobs in the industry and services sector. Women more often have a part-time job than men. In intermediate and most urban regions about one-third of employed women have a part-time job. In the group of the most rural regions only one-sixth of all employed women are working part-time. Especially in the lagging most rural regions, the share of women with part-time jobs is low.

Table 3.14 Share of part-time workers in total employment, 1991[a]

	Total	Agriculture	Industry	Services	Female part-time jobs in total female employment
Leading most rural	0.11	0.15	0.06	0.12	0.24
Leading intermediate	0.14	0.21	0.07	0.19	0.30
Lagging most rural	0.07	0.11	0.03	0.06	0.12
Lagging intermediate	0.14	0.12	0.06	0.21	0.29
Most rural regions	0.08	0.12	0.04	0.08	0.16
Intermediate regions	0.15	0.16	0.06	0.20	0.32
Most urban regions	0.17	0.17	0.07	0.22	0.35
EU average	0.16	0.15	0.06	0.20	0.32

a) Analysis is based on data for Belgium, Germany, Spain, Ireland, Italy, Luxembourg, Netherlands and Finland and on some regions from Greece, Portugal and Sweden.

Employment growth and farmers with other gainful activity: no relationship

Some information about the characteristics of the agricultural labour force can be useful to understand the features of the labour market in rural regions. The highest rate of part-time farmers is found in the group of intermediate regions (67%), whereas it is at a lower and about equal level in the group of most rural and most urban regions (59 and 57%) (Table 3.15). However, among countries the percentage of part-time farmers fluctuates from 32% in the Netherlands to about 80% or more in Portugal, Spain, Greece and Italy. The percentage of farmers with other gainful activities falls from 36% in the group of most rural regions to 31% in the group of most urban regions. It could be expected that the percentage of farmers with other gainful activities is affected by employment growth as this offers opportunities for other gainful activities. However, a relationship between employment growth and the percentage of farmers with other gainful activities does not emerge from our data (Table 3.15).

Higher share of young farmers in most rural regions

Regarding the farmers' age structure, it appears that most rural regions have the highest presence of young farmers and the lowest of old farmers; the opposite behaviour can be observed in urban regions although the differences with intermediate ones are very small (Table 3.15). One explanation of this result can be the different role played by agriculture. In urban regions, agriculture is a marginal sector from the point of view of employment, and farmers tend to be workers without valid external alternatives; the young find alternative job opportunities outside the agricultural sector more easily. In rural regions these

Table 3.15 Part-time farmers, farmers with other gainful activity (OGA), farm age structure and correlation OGA with employment growth, 1989/90 (%)

	Part-time[a]	OGA[b]	<35 year[b]	≥ 65 year[b]	Correlation coefficients OGA and non-agr. empl. growth[b,c]
Leading most rural	61	33	11	19	-0.08
Leading intermediate	66	33	9	20	-0.31
Lagging most rural	55	35	11	19	0.26
Lagging intermediate	67	31	8	23	-0.21
Most rural regions	59	36	12	18	-0.04
Intermediate regions	67	33	9	21	-0.07
Most urban regions	57	31	8	22	0.02
EU average	61	28	9	21	-0.03

a) Data is available for all countries, except for Austria and Sweden and for some regions in Denmark and the UK; b) Data is available for all countries, except for some regions in Denmark and the UK; c) Correlation between the percentage of farmers with other gainful activities and (relative) non-agriculture employment growth.

alternatives are less abundant, and agriculture continues to play a central role for the employment of the young labour force.

Concluding remarks

In this chapter we have examined some socio-economic characteristics of leading and lagging regions in the EU. For this purpose, we have constructed various clusters with regions from different EU countries. Our database has some shortcomings for carrying out an EU-wide analysis, which mainly refer to lack of data for some variables and differences in national definitions for variables among countries. These shortcomings limit the results of our analysis to some extent. Besides, differences in the level of some variables exist among countries, like GDP per capita, unemployment rates and participation rates. These differences are related to structural and country specific factors. As we have not in all cases corrected for these factors, a country bias in the results may arise.

Seven out of the 11 hypotheses examined in our study were supported by our analysis. These are:

- Rural regions with employment growth tend to have population growth as well.

- Employment growth tends to be higher in rural regions with a higher GVA/worker (only for most rural regions).
- Rural regions with a higher endowment of infrastructure tend to show a higher employment growth.
- In rural regions the education level of the population is lower than in urban regions.
- In rural regions with a higher education level of the population, the economy is more diversified than in other rural regions.
- Rural regions with high unemployment rates tend to have lower participation rates than rural regions with low unemployment rates.
- In rural regions with a higher education level of the population, participation rates tend to be higher.

The other hypotheses were not supported by the analysis. These hypotheses are:

- Employment growth tends to be higher in rural regions with a higher GVA/worker.
- In rural regions with a higher education level of the population, the increase in employment is higher than in other rural regions.
- Rural regions with employment growth tend to have higher participation rates than rural regions in which employment growth stagnates.
- The percentage of farmers with other gainful activities in rural regions with employment growth tends to be higher than in rural regions in which employment growth stagnates.

References

Esposti, R., Godeschalk, F.E., Kuhmonen, T., Post, J.H., Sotte, F. and Terluin, I.J. (1999) *Employment growth in rural regions of the EU; A quantitative analysis for the period 1980-1995.* LEI, The Hague.

OECD (1994) *Creating rural indicators for shaping territorial policy.* Paris.

OECD (1996) *Territorial indicators of employment; Focusing on rural development.* Paris.

Employment Dynamics in Two Neighbouring Rural Regions of Belgium and France

4

Bruno Henry de Frahan, Pierre Dupraz and Béatrice Van Haeperen

Introduction

The two neighbouring rural regions contrasted in this chapter are the province of Luxembourg located in southeast Belgium and the Département Les Ardennes located in northeast France (Fig. 4.1)[1]. Both regions have similar area and population sizes and a low population density of about 55 inhabitants per square kilometre, with no towns greater than 30,000 inhabitants except for Charlevilles in Les Ardennes (Table 4.1). Despite sharing some common geographical, historical and economic development features, they have significantly diverged in terms of gross domestic product (GDP) and employment growth during the last two decades. They have also significantly diverged from the GDP and employment growth of their mother country. While Luxembourg (B) was outperforming the GDP and employment growth of its country, Les Ardennes was far below the national level of GDP and employment growth in its own country (Table 4.2).

Luxembourg (B) provides an interesting case to compare with the neighbouring Département Les Ardennes. Luxembourg (B) is the only NUTS2 predominantly rural region in Belgium according to the OECD definition (OECD, 1994)[2]. With 8% of its labour force in the agricultural sector, it is the most agricultural region in Belgium. Luxembourg (B) is also the region of the country where population and employment grew at the fastest rate between 1980 and 1993. In 1995, the unemployment rate in the province was 9% compared with 12% in Belgium. Although the number of its residents commuting to the neighbouring Grand Duché du Luxembourg to work grew from 6,000 in 1981 to 13,000 in 1994, Luxembourg (B) can nevertheless be considered as a leading region in terms of employment growth: the number of employed persons whose

---

[1] In this chapter, the Belgian province of Luxembourg is abbreviated by Luxembourg (B) and the French Département Les Ardennes by Les Ardennes.
[2] The provinces of Liège and Namur are intermediate regions, while the other seven provinces of the country are predominantly urban regions.

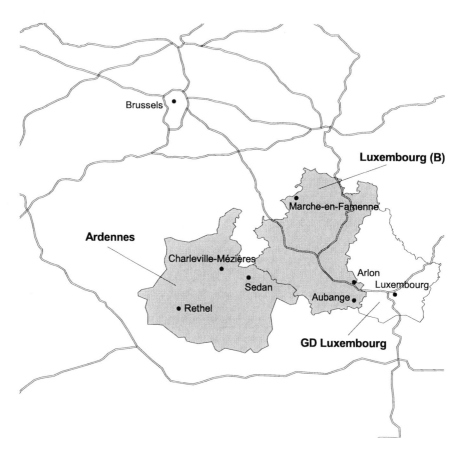

Figure 4.1   Main cities and roads in Luxembourg (B) and Les Ardennes

place of work is located in the province, grew from 66,300 in 1981 to 80,800 in 1994, i.e. a rate of net employment creation within the province of 22% compared with 5% for the country[3]. This exceptionally high employment growth performance provides an interesting case to study.

Les Ardennes is one of the 96 NUTS3 predominantly rural regions of France and, like Luxembourg (B), has 8% of its labour force in agriculture. Population and employment in Les Ardennes decreased at a higher rate than in the group of rural regions in France: respectively by 6% and 3% in Les Ardennes compared with 4% and 1% in the cluster of predominantly rural regions between 1980 and 1992. In 1995, the unemployment rate in Les Ardennes was 15% compared

---

[3] Total employed population being resident in Luxembourg (B), including the commuters, increased by 30% during the same period.

Table 4.1 Demographic, geographic and economic indicators for Luxembourg (B) and Les Ardennes and their mother countries

	Luxembourg (B)	Les Ardennes	Belgium	France
Population, 1993	237,500	293,500	10,084,500	57,529,300
Population growth, 1980-1993 (% p.a.)	0.5	-0.2	0.2	0.5
Size (km²)	4,440	5,250	30,530	544,000
Population density, 1993 (inh./km²)	54	56	332	105
Population largest town, 1990	23,300	67,200	950,000[b]	9,000,000[c]
Number of municipalities	44	421	589	36,551
Agricultural area (ha)[a]	137,300	332,900	1,345,000	30,060,000
Forest area (ha)[a]	207,900	152,600	610,000	16,050,000
Number of farms[a]	5,300	4,300	81,000	735,000
Nominal GDP growth, 1980-1993 (% p.a.)	6.4	5.3	5.9	6.4
GDP per capita, 1993 (ECU)	14,300	15,900	17,900	18,500
Unemployment rate, 1985 (%)	10	14	11	10
Labour force growth, 1980-1995 (% p.a.)	1.4	-0.25	0.5	0.5
Non-agricultural employment growth, 1980-1992 (% p.a.)[d]	1.2	- 0.4	0.2	0.4

Sources: De Schrevel (1994), EUROSTAT, INS and INSEE.
a) 1992 for Luxembourg (B) and 1995 for Les Ardennes; b) Agglomeration of Brussels (136,400 dwellers without suburbs); c) Agglomeration of Paris (2,152,400 dwellers without suburbs); d) 1981-1992 for Les Ardennes and France.

with 13% in France. Due to its low employment performance, Les Ardennes can be considered as a lagging region in terms of employment. Since both Les Ardennes and the neighbouring province of Luxembourg were confronted with a restructuring of their declining iron and steel industries in the 1970s, the relatively low employment growth performance in the lagging region provides an interesting case to compare with the leading region of Luxembourg (B).

Local resources

The forest plateau of the Ardennes crosses both the province of Luxembourg and Les Ardennes. The fringes of the Ardennes forest have been traditionally devoted

Table 4.2 Growth of employment and nominal GDP, 1980-1992 (% p.a.)

	Total employment	Non-agricultural employment	GDP
Belgium			
Luxembourg (B) (predominantly rural)	0.9	1.2	6.4
Liège (intermediate rural)	-0.5	-0.5	5.5
Namur (intermediate rural)	0.2	0.2	5.9
Predominantly urban provinces	0.2	0.2	6.0
Belgium	0.1	0.2	5.9
France [a]			
Les Ardennes (predominantly rural)	-0.6	-0.4	5.3
Predominantly rural Départements	-0.1	0.5	6.1
Intermediate rural Départements	0.4	0.8	6.2
Predominantly urban Départements	0.2	0.3	6.7
France	0.1	0.4	6.4

Source: Eurostat.
a) 1981-1992.

to meadows and pastures while flat areas have been cultivated. On and around this plateau, both regions have specialized in forestry products and livestock breeding and, to a greater extent in Les Ardennes, in milk production. The large fertile plain located in the southern part of Les Ardennes has been cultivated with cereals, sugar beet and industrial crops, which stimulated until recently a local processing industry. Both regions have countryside and historical monuments and sites equally suitable for tourism and recreation. Differences between the two regions in land endowment, communication network as well as population growth and distribution affected the development path of their economic activities and employment.

Land endowment

One main resource in both rural regions is land that not only is abundant and cheap for economic development, but also provides a quiet non-congested environment and green landscape for enterprise and people settlement. This resource is, however, more valuable in Luxembourg (B) than in Les Ardennes. Being the only rural region in Belgium, Luxembourg (B) is relatively better endowed in this resource than the other regions of the country, while Les Ardennes is not better endowed in this resource than the other regions in the northeastern part of France. Greater and faster road improvements in Luxembourg (B) than in Les Ardennes also facilitated the access to this resource. The comparative advantage in this resource has, therefore, been greater in Luxembourg (B) than in Les Ardennes. As a result, the quiet environment and green landscape associated with land availability made Luxembourg (B) relatively more attractive in the Belgian context for people and enterprises. Responding to growing labour demand in the neighbouring Grand Duché du Luxembourg, people immigrated to Luxembourg (B) to take advantage of this resource which was relatively more abundant and cheaper than in the Grand Duché du Luxembourg. Large foreign enterprises settled in Luxembourg (B) to take advantage of the relatively low land price and space amenities. In contrast, Les Ardennes did not exploit its comparative advantage in land to the same extent as Luxembourg (B).

The communication network

Communication networks are helpful in supporting economic development in rural areas, particularly in industry and trade services where 'just in time' practices are crucial. Differences in relative communication network endowments between the two regions are striking (Fig. 4.1). In Luxembourg (B), the road and motorway networks were remarkably improved between 1980 to 1995 as a result of the active lobbying of politicians from the province at the national and European levels. For instance, the motorway length increased from 7 to 154 km. It is now less than a 2-hour drive to reach Brussels, Liège or Luxembourg city from anywhere in the province. Several railway lines cross the province, in particular in the south of the province. Except for motorways, Luxembourg (B) is still less equipped with roads and railways than the other provinces of the country.

In Les Ardennes, the waterway and railway were key instrumental assets in the former development of the heavy iron and steel industry that linked this industrial basin with the other mining and industrial basins in the north and northeast of France. With the decline of this industry, this network has lost its strategic importance in the economic development of the Département. Since then, the economic development of Les Ardennes has been handicapped by weak road connections with its neighbouring regions. Of the main roads connecting the centre of the Département in the southern direction with France and in the northern and eastern direction with Belgium, only 13 kilometres are motorway. As a result, despite its favourable geographical location in the middle of northwest Europe, Les Ardennes is not directly connected to the motorway network of this large and rich area. Unlike Luxembourg (B), Les Ardennes is not crossed by the main flows of commodities and tourists. Poorly co-ordinated actions at the national and European levels from the regional authorities partly explain this weaker communication network endowment of Les Ardennes.

Population growth and distribution

Population growth has a double impact on the local economy, first through its impact on the demand for services and, second, through its impact on the supply of labour. On the other hand, spatial population distribution affects the spatial integration of the economic activities and influences regional development strategies. Population growth and distribution in Luxembourg (B) and Les Ardennes as well as their development strategies have been dissimilar.

Between 1980 and 1993 the population increased in Luxembourg (B) but decreased in Les Ardennes. Accordingly, the labour force, as a measure for the labour supply, increased in Luxembourg (B) but decreased in Les Ardennes (Table 4.1). Population migration, not natural growth, explains this opposite pattern. In Luxembourg (B), net immigration became stronger, beginning in 1987 with the revival of the economic growth in the province, then was at its maximum in the early 1990s with the completion of the new motorway between Brussels and Luxembourg city. Half of the immigrants came from more congested urban regions of the country while the other half came either from other EU countries (30%) and non-EU countries (20%). In Les Ardennes, net emigration started in the 1960s with the collapse of the textile industry and the increasing number of young educated people leaving the region. In 1975 the population started to decline and, since then, the service sector of the Département has hardly grown.

Population is more evenly spread in Luxembourg (B) than in Les Ardennes. The population density has, however, been traditionally higher in the two southern districts of the province of Luxembourg, particularly in the district of Arlon close to the Grand Duché du Luxembourg where the iron and steel industry developed. In the future, population growth is expected to be higher in the three northern districts of the province. Because of the attraction of the Grand Duché du Luxembourg for job opportunities, Luxembourg (B) is hardly a closed functional labour market. In contrast, the population and economic activities of Les Ardennes have been traditionally concentrated in the northern urban municipalities along the Meuse river where the iron and steel industry developed. About 60% of the population and most of unemployed population inhabit this industrialized area that is smaller than 15% of the area of the Département. This area is the main functional labour market of the Département. Mostly dedicated to agriculture, the southern area of the Département is much less inhabited. As a result of the decline of the food industry and the improvement of the main north-south road, this southern area has increasingly become a part of the labour market of Reims, which is an economic centre of about 200,000 inhabitants in the neighbouring Département of Marne.

A more even spatial distribution of the population in Luxembourg (B) may explain the reason why the regional authorities of Luxembourg (B) applied for different European programmes with a view to cover the whole population of the province. The Objective 2 and 5b programmes have respectively covered the southern district of Arlon and the northern districts of the province, while the European LEADER initiative has covered several rural municipalities across the province. In contrast, the regional authorities of Les Ardennes only applied for the Objective 2 programme although a large part of the Département was also eligible for the Objective 5b programme. As a result, very little has been done in the Département to valorize the rural amenities and exploit them in a way to increase employment.

Economic activities

Both regions saw their traditional iron and steel industry modernized and expanded between 1950 and 1970 with workers being also small farmers. This expansion was concentrated in the extreme South of Luxembourg (B) and, to a larger extent, along the Meuse river in the northern part of Les Ardennes. Along this river, the automotive industry was simultaneously developed. The traditional textile industry located in the eastern part of Les Ardennes enjoyed a short revival during the same period. In the late 1970s, the European crisis in the iron and steel industry struck both regions. In Luxembourg (B), the southern iron and steel site was abruptly closed in 1977 with its remaining 1,665 workers put out of work. In Les Ardennes, the iron and steel crisis started to affect employment beginning in the early 1980s and extending over a much longer period than in Luxembourg (B). The crisis culminated in 1983 with the closure of two large state iron and steel plants along the Meuse river that put about 5,000 workers out of work, in addition to those who became unemployed due to the closure of several family-run medium-size sub-contracting firms, which lacked the skills and capital for conversion. Hit by competition from artificial textile products, the once flourishing textile industry of Les Ardennes disappeared in the 1970s and, consequently, many unskilled female workers were out of work. As a result, the manufacturing industry of Les Ardennes lost about 8,500 jobs between 1981 and 1993. More recently, most of the food processing plants in the southern part of Les Ardennes have been closed as a result of the concentration of the processing activities in this economic sector. These activities have been merged in plants located in neighbouring regions.

On both sides of the border, restructuring was unavoidable in the iron and steel industry as well as in the Les Ardennes' textile industry (Henry de Frahan *et al.*, 2000). To compensate for the employment loss in the iron and steel industry, the public administration of Luxembourg (B), headed by its dynamic governor, reacted and, together with local business support agencies, succeeded in attracting a dozen subsidiaries from multinational companies. These companies were offered not only fiscal incentives and investment subsidies that eventually amounted up to 20% of the total investment, but also services to facilitate their setting up as well as the arrival and settling in of the foreign managers and their families[4]. This concerted action generated more than 2,000 jobs between 1977 and 1990. In the mid-1980s, the governor of the province of Luxembourg, with the prime minister of the Grand Duché de Luxembourg and the head of the French Region of the Lorraine, initiated a multi-country European Development Pole (EDP) with the objective of generating 8,000 new jobs on the former site of the iron and steel industry of the three bordering regions. Due to national and European investment subsidies amounting here to about 30% of the total

---

[4] The 1970 national law of economic expansion was restricted to enterprises employing more than 250 workers with a turnover above 20 million ECU, i.e., mostly subsidiaries from foreign multinationals located in a identified 'development area' while the 1978 national law of economic reorientation applied to enterprises employing less than 250 workers with a turnover below 20 million ECU without area restriction.

investment and the active role of local business support agencies, 1,450 more jobs were generated between 1988 and 1995 on the Belgian side of the former industrial site. In the meantime, the provincial lobbies were able to have the road network dramatically improved connecting the province to the main European economic centres since 1990. This restructuring as well as the improvement of the overall economic situation helped the revival of the provincial economy since 1987, with the private sector taking over from the public sector in employment creation.

In Les Ardennes, some existing small and medium-size enterprises (SME) of the iron and steel industry were able to adapt to the industrial crisis and resist competition. These SME used national assistance funds either to specialize in some metallurgy, foundry and steel works or to diversify in intermediate goods for the automotive and other industries. Some other SME were bought by larger ones or disappeared. In the late 1980s, several high technology subsidiaries from multinational automotive companies specializing in plastics also settled in Les Ardennes, thanks to national investment subsidies amounting to 12% of the total investment[5]. These large plants form now a dynamic capital-intensive industrial centre of plastic manufacturing employing some 6,000 workers. Because they did not have a unified territorial strategy towards employment, the public administration of Les Ardennes, however, hardly exploited the available European recovery programmes for creating jobs.

Employment and labour productivity growth

During the studied period of 1980-92, employment grew at an annual rate of 0.9% in Luxembourg (B), but declined at an annual rate of 0.6% in Les Ardennes (Table 4.2). With respect to the intermediate and urban regions of Belgium, employment growth in Luxembourg (B) during that period was exceptionally high (Fig. 4.2), whereas employment decline in Les Ardennes was worse than in the clusters of the rural and urban regions of France. Made up of an urban area along the industrialized Meuse river and deep rural areas elsewhere, Les Ardennes seems to have cumulated the unfavourable features of both these types of areas for employment growth since the cluster of the intermediate regions of France was the only one to show a growth in employment. Labour productivity, however, grew at a similar annual rate of 6% in both rural regions[6]. This rate is also similar in the other clusters of regions of both Belgium and France, except in the cluster of the urban regions of France, which had a rate of almost 7%. Given an employment growth in Luxembourg (B) but an employment decline in Les Ardennes and a labour productivity growth similar in both regions, the regional economy grew slightly faster in Luxembourg (B) than in Les Ardennes (Table 4.1).

---

[5] In addition to subsidies for modernizing existing small enterprises, national subsidies were available for investments over 3 million ECU to create more than 20 full-time jobs for enterprise creation and 50 full-time jobs for enterprise enlargement. Depending on the location, combined European and national investment subsidies reached 17 to 25% of total new investment and 27% for the modernization of SME.

[6] Labour productivity is measured as the per worker gross value added at current prices.

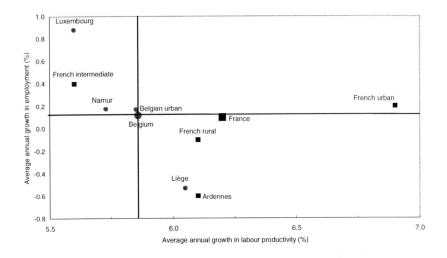

Figure 4.2   Growth of employment and labour productivity (at current prices) in the regions of Belgium and France, 1980-1992 (% p.a.)
Source: Derived from Eurostat.

Employment growth and sector mix

Part of the high employment performance of Luxembourg (B) with respect to Les Ardennes can be explained by the difference in their sector mix. We know that, in post-industrialized countries, the service sector generates most of the employment growth, while agriculture and industry generally experience an employment decline as a result of a substitution of labour by capital. In the early 1990s, the economy and employment of Les Ardennes continued to depend more on industry than Luxembourg (B) (Table 4.3). Being particularly capital-intensive, industry in Les Ardennes could not generate employment. In contrast, the economy and employment of Luxembourg (B) depended more on the labour-intensive service sector than in Les Ardennes.

During the restructuring period, due to labour productivity increase the rate of farm enlargement was faster in Les Ardennes than in Luxembourg (B). As a result, employment loss in the agricultural and forestry sector was greater in Les Ardennes than in Luxembourg (B), respectively 30% and 11% (Table 4.4 and Fig. 4.3). Luxembourg (B) was able to increase its employment in the manufacturing branch by 12% while Les Ardennes lost 24% of its jobs. Both Les Ardennes and Luxembourg (B) lost jobs in the construction branch, respectively 20% and 27%. Employment in the different branches of the service sector increased much more in Luxembourg (B) than in Les Ardennes, respectively 31% and 6%. Between 1985 and 1995, the total unemployment rate declined from 9.5% to 7.0% in Luxembourg (B), but increased from 13.9% to 14.4% in Les Ardennes.

Table 4.3 Sectoral shares in GDP and employment (%) and GVA per worker (ECU) in Luxembourg (B) and Les Ardennes, 1992-1994

	Luxembourg (B)			Les Ardennes		
	Share GDP (1993)	Share employment (1992)	GVA per worker (1992)	Share GDP (1994)	Share employment (1992)	GVA per worker (1992)
Agriculture	7	8	35,300	4	8	26,810
Industry	18	19	41,030	39	35	50,850
Services	75	73	41,740	57	57	45,400
Total	100	100	41,150	100	100	45,900

Source: Eurostat.

Table 4.4 Employment change across sectors in Luxembourg (B) and Les Ardennes, 1981-1993

	Luxembourg (B)				Les Ardennes			
	Total employed		Change 1981-1993		Total employed		Change 1981-1993	
	1981	1993	abso-lute	rela-tive (%)	1981	1993	abso-lute	rela-tive (%)
Agriculture and forestry	8,888	7,879	-1,009	-11	10,500	7,338	-3,162	-30
Manufacturing	9,893	11,035	1,142	12	35,473	26,973	-8,500	-24
Construction	6,039	4,439	-1,600	-27	6,736	5,369	-1,367	-20
Trade, hostel, restaurant, repairs	11,723	14,712	2,989	26	18,218	16,108	-2,110	-12
Transportation, telecommunication	3,390	4,583	1,194	35	5,547	4,715	-832	-15
Finance, insurance, house sale and renting	1,380	3,150	1,770	128	1,801	2,117	316	18
Other private services	2,740	4,285	1,545	56	9,201	12,930	3,729	41
Non-market services	23,349	29,045	5,696	24	17,688	19,680	1,992	11
All activities	67,402	79,127	11,726	17	105,164	95,230	-9,934	-9

Source: INASTI and ONSS for Luxembourg; INSEE for Les Ardennes.

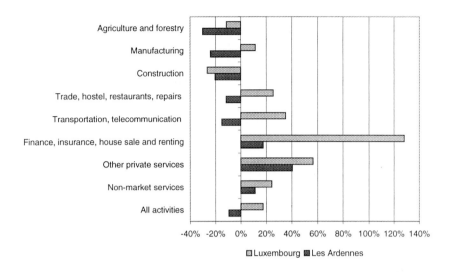

Figure 4.3   Employment change across sectors in Luxembourg (B) and Les
Ardennes, 1981-1993 (%)
Source: INASTI and ONSS for Luxembourg; INSEE for Les Ardennes.

Less aggregated data show that about half of the net increase in jobs actually
occurred in public services (administration, education, health, entertainment, etc.)
in Luxembourg (B) and one-third in Les Ardennes. A significant net increase in
jobs occurred in other branches also characterized by limited exposure to global
markets, relatively stable prices and rather labour intensive production: trade and
tourism in Luxembourg (B), and other private services to households and
enterprises in both regions. Only in Luxembourg (B), there was a significant net
increase in jobs in branches exposed to global markets and price fluctuations:
manufacturing, transport and communication, and financial services. The job
losses in Les Ardennes occurred not only in the manufacturing sector but also in
the trade, transport and communication sectors. Compared with Luxembourg (B),
job losses in Les Ardennes were fewer in the construction sector as a result of its
lower labour cost.

The public sector apart, new firms contributed mostly to job creation in both
regions. In Luxembourg (B), enterprises of either fewer than 10 employees or
more than 100 employees created relatively more employment than medium size
enterprises. In Les Ardennes, small but also medium-size enterprises created
more employment while large manufacturing enterprises experienced most of the
employment loss. The labour participation rate, female share of the labour force
and the share of part-time jobs in total jobs increased similarly in both regions
during the restructuring period.

Regional leadership and employment strategies

Our study of the province of Luxembourg and Les Ardennes revealed that the ability of local leaders such as regional authorities and administrators, heads of professional associations and business support agencies, and trade union representatives to implement conditions for successful territorial strategies, is a key factor in explaining the difference in employment growth of the two rural regions (Henry de Frahan *et al.,* 2000). The organization and role of local leaders as well as other actors such as entrepreneurs have already been underlined in the growing field of economic research on industrial districts as surveyed in Benko *et al.,* (1997) and Barré (1998). The industrial district is a system of SME organized locally for the supply of both goods and services. The development of these industrial districts is generally endogenous, based on the capacity of the local actors rather than on the local biophysical resources, and rooted in an institutional and political environment. As shown below, Luxembourg (B) shares several characteristics of industrial districts such as a strong local identity, a strong consensus on the defence of collective economic interests among local leaders, a dense network of associations and representatives of actors, while Les Ardennes does not.

Effective endogenous and territorial strategy in Luxembourg (B)

In Luxembourg (B), the collapse of the iron and steel industry in 1977 spurred a strong territorially-based strategy of economic and employment development. At that time, the leaders of the province had a defensive strategy: their main goal was to replace the 1,700 jobs lost in the steel and iron industry and avoid investments in a single industrial branch. To do this, they tried to attract foreign investments in the province, and were successful. Even if the role of exogenous forces in recovering the previous employment level was essential, this process actually was the result of a strategy co-ordinated at the level of the local leaders – a bottom-up strategy – and so, endogenously determined. Among the factors explaining the success of this strategy, we find a strong political and social will among local leaders to defend collective economic interests, a strong local identity and the involvement of active policy makers and other local leaders in the development process.

Thanks to this general political and social consensus, local leaders were able to stimulate development and employment growth in the province by creating or enhancing the conditions that favour enterprise settlements and endogenous development. They encouraged the institutional development of organizations representing local actors: the municipality syndicate (IDELUX), the provincial chamber of commerce and industry (CCILB) and the regional interest group *Fondation rurale de Wallonnie* (FRW)[7]. The strong co-operation between the provincial authorities and IDELUX during the restructuring period stimulated co-operation among other institutions that, elsewhere, are prone to competition or

---

[7] Local and national policy makers and university professors created the FRW to defend and promote the economic interests of the province at higher institutional levels.

discord: municipalities, the provincial administration, the public employment and vocational training institution (FOREM) and the training institution for SME (IFPME), trade unions and business organizations. This cooperation avoided the dispersion of the efforts, facilitated the exchange of information and services and diffused social conflicts to the point that strikes have been avoided. Thanks to this cooperation, public administrations have been able to anticipate the needs of the firms of the province by finding and lobbying national and European subsidies for them, orienting professional training to their needs and informing them about labour availability. The effective cooperation among the public and private institutions of the province enabled a considerable reinforcement of networks of local actors. These strongly cooperative internal networks were the vehicle for a new positive image of the region which helped attract new firms to the province. Moreover, lobbying activity initiated by the FRW and supported by the provincial authorities, the aim of which was to achieve parity in financial support between rural and urban regions, proved to be very successful.

This well-developed networking proved to be an efficient way to understand the problems encountered by the enterprises, to respond properly to these problems and to diffuse information. One problem encountered by most of the local enterprises in Luxembourg (B) that became prominent after the restructuring period, was their small size. Due to insufficient managerial skills, lack of time for acquiring managerial training, risk-averse attitude towards investment and high administrative costs to collect valuable information, these small enterprises hardly expanded beyond five employees. To overcome these difficulties, the municipality syndicate and the provincial chamber of commerce and industry have been involved in advising and training SME managers. Local leaders have since refocused their economic and employment strategy to create or enhance the socio-economic conditions favouring endogenous development. To encourage further investments in both manufacturing and service sectors, priorities have recently shifted to supply services to small enterprises to improve their managerial skills as well as horizontal and vertical co-ordination among local firms, while continuing to emphasize the regional comparative advantage, like the quiet and green environment of the province.

External networks developed considerably both in the public and private spheres at the regional, national and European levels. Political and economic leaders from the province usually worked together in networking activity. Consequently, the provincial authorities were able to negotiate with the national and regional authorities on the basis of a political consensus reached at the provincial level, particularly for the improvement of infrastructure such as the road network and for the land-use plan, which determines new sites for enterprises. They were able to impose a territorial approach to solve their problems and exploit fully regional, national and European investment and employment measures. Among them, the most frequently used measures included investment subsidies with their associated premium for employment creation. The provincial authorities were also able to innovate by elaborating and negotiating subsidy packages from national or European sources on a case by case basis as well as by developing a positive image of the province symbolized in a digitized green wild boar representing enthusiasm for work, modernity and greenness of the province. The strong presence of subsidiary businesses in the province, the multi-country EDP scheme strengthening links among local actors of the province, the Grand Duché de Luxembourg and the French Lorraine, as well as other cross-border initiatives provided the province with many economic opportunities.

An ineffective competitiveness-based economic strategy in Les Ardennes

In Les Ardennes, the main goal was to foster regional competitiveness and economic growth, but without a specific focus on employment growth. A large part of new investments was in capital intensive high-technology industries – automotive and plastic industries. This choice was mainly made by large enterprises, rather than resulting from a regional consensus. Some measures designed at the national level for regions facing industrial restructuring were used in Les Ardennes. However, due to a lack of leadership in Les Ardennes, ineffective lobbying at the national or European levels and poor internal organization, the available subsidies did not give rise to a general and coordinated territorial strategy. In the agricultural sector, local authorities and farm organizations hardly addressed the employment loss resulting from the concentration process. In the tourist sector, local authorities decided not to increase the number of enterprises in order to protect the profitability of the existing ones. Hence, employment in this sector did not increase.

Compared with Luxembourg (B), Les Ardennes proved incapable of designing an effective territorial strategy of development for several reasons. First, weak and scattered representative organizations and insufficient networking activities among them were a handicap in attaining consensus about regional development and may, at least partially, explain this difference. Second, there was no political consensus among local authorities. One reason may be the high number of municipalities, which is ten times the number of municipalities in Luxembourg (B) for a population and area of the same size. Several groups of municipalities tried to create associations or syndicates for different purposes, but no actual territorial policy has emerged. Third, Les Ardennes had no political influence at the national level, and the local entrepreneurs and policy makers were unfamiliar with European structural programmes. This impeded the completion of road infrastructure and full exploitation of measures offered at the national and European levels for fostering employment. For example, the highway network that the Département started in the 1960s has not yet been completed. The resulting poor road connections between the main economic centres are considered by many companies specializing in intermediate goods as a major handicap. To cite another example, while European structural funds for rural areas were available, the authorities of Les Ardennes decided not to apply for these funds, opting for European structural funds for industry restructuring instead. Moreover, the poor coordination between local leaders resulted in insufficient efforts to provide outside potential investors with customized aid packages, along the line of those provided in Luxembourg (B). As a result, the available European structural funds were not fully used. It is only in the manufacturing sector that external networks were instrumental in modernizing and restoring the competitiveness of the region. This modernization ensured the economic sustainability of the remaining jobs but at a cost of employment loss.

In sum, local leaders in Luxembourg (B) could rely on a strong regional identity and political and social consensus and were able to exploit their formal and informal internal as well as external networks to design and implement a territorial development and employment strategy. This territorial strategy aimed first to attract many public funds and foreign direct investments and, more recently, to support the start-up of new SME and the development of existing ones. Thanks to a handful of dynamic and dedicated leaders, including the governor of the province, this territorial strategy was successful in reversing the decline in labour employment despite the high risk aversion and low innovation capacity of the local entrepreneurs. In contrast, local leaders in Les Ardennes could not rely on a political and social consensus and strong internal as well as external networks. Hence, they were unable to agree on a common development and employment strategy and act together. European structural funds were under-utilized. The local entrepreneurs of the surviving companies progressively adapted to market and technological changes but met difficulties in research and development as well as marketing activities as a result of their small size. In both cases, despite their positive attitude towards their work, workers and employees could not provide the education and skill levels that was demanded.

Key factors in employment dynamics

A favourable sector mix

Multiple factors made employment growth easier in Luxembourg (B) than in Les Ardennes (Dupraz *et al.*, 1998a, b). The manufacturing sector in need of restructuring at the end of the 1970s was a smaller part of the regional economy and was concentrated among fewer firms and in a smaller area in Luxembourg (B) than in Les Ardennes. With a sector mix already largely diversified into services, restructuring was therefore more manageable in Luxembourg (B) than in Les Ardennes.

A favourable location

The proximity of a growing economy localized in the neighbouring Grand Duché de Luxembourg in need of labour for its dynamic manufacturing and service sectors was equally favourable for Luxembourg (B). Not only did this across-the-border dynamic economic area absorb some extra labour from Luxembourg (B), but it also stimulated immigration in Luxembourg (B) from other parts of Belgium. Because of a growing population, private and public services in Luxembourg (B) continued to grow and hire labour. As a result, the development of services in Luxembourg (B) could balance the employment loss in the other economic sectors. In contrast, unemployed labour in Les Ardennes was attracted to large economic centres like Reims or Paris and settled there instead of choosing daily commuting from Les Ardennes, because of the distance, but also because of the better services and infrastructure in these larger cities. The population decline in Les Ardennes slowed the development of services. As a result, employment in services could not match the employment loss in the other economic sectors.

Effective regional institutions

A regional institutional framework concentrated among fewer municipalities and business supporting agencies was also favourable to Luxembourg (B) in the sense that it could design and implement an effective territorial approach to employment growth. This territorial approach to employment growth has been facilitated by several factors. The unique characteristics of Luxembourg (B) with respect to the other provinces of the country in terms of its relative remoteness, poverty and development fostered a strong cultural and social identity that was able to unite local leaders from different social and political backgrounds for concerted political actions. These local leaders were successful in making this uniqueness recognized at the national and European levels and in accessing subsidized public funds. Clever financial engineering and active marketing towards foreign private investors permitted an extensive use of the available regional, national and European public funds for economic development. Space availability, the quiet environment, the green landscape, easy connections to the European road and railway networks, a hard working labour force and social peace were all assets of the province that were emphasized by the local leaders and business agencies towards potential external investors.

In contrast, the large number of municipalities grouped into several associations and syndicates of different purposes was not favourable to Les Ardennes to design, agree upon and implement an effective strategy towards employment growth. The relative remoteness, poverty and development of Les Ardennes are regional characteristics that are less unique in France than Belgium. These geographic and economic characteristics could not foster a cultural and social identity strong enough to overcome rivalries between local leaders and organizations and reach a well-designed consensual employment strategy similar to the strategy in Luxembourg (B). The relatively skilled and cheap manufacturing labour force and the availability of national and European structural funds in Les Ardennes were assets that local leaders could not fully exploit to attract potential investors as a result of a poor coordination and political will among them.

An unified employment strategy

The unified development and employment strategy developed by the political and economic leaders of Luxembourg (B) rested both on endogenous and exogenous forces. To address the under-development of their region and the loss of employment associated with the iron and steel industry restructuring in the 1970s, the local policy makers were clearly the initiators of an employment replacement strategy that included building and exploiting strong external networks to claim territorial parity and even positive discrimination in public investment allocations from the regional and national levels and attract private investments from outside. Through internal networks they shaped and realized an ambitious policy of economic and territorial development by ensuring a political and social consensus and building a renewed image of the region. Exogenous forces at work were settlement of foreign enterprises or subsidiary businesses with managers and private and public investment coming from the outside and the improvement of the road, highway and telecommunication networks which opened the region to space consuming activities such as large plants, military settlements, short stay tourism and outdoor recreation activities. Specific policies instrumental to development and employment growth consisted of improving the infrastructure (road, water, energy and sanitation), consolidating and coordinating the new municipalities in the development agency of IDELUX, implementing the European Development Pole which provided investment aid for new firms and infrastructure and preparing and equipping industrial sites. Subsidy packages for attracting new firms were individually negotiated and provided. Entrepreneurs did not develop specific strategies towards maintaining or augmenting employment. To maintain a favourable social climate, they preferred to negotiate internal flexibility such as work sharing and temporary lay-off to firing. Labourers accepted part-time and commuting work, the increase of which explains most of the employment growth in the region. Self-employment is another important source of employment.

In contrast, Les Ardennes clearly suffered from a lack of leadership and unified strategy due to the economic heterogeneity of the region and lack of solidarity between the different parts of the region. Forces exogenous to the region mainly explain the regional development and employment dynamics. Global market forces were responsible for the decrease in employment in the manufacturing and the agricultural sectors as a result of the iron, steel and textile crisis and farm enlargement respectively as well as the resulting emigration of young educated people. After the decline of the iron, steel and textile industry, policy makers only applied for EU support in the Objective 2 area but not in the Objective 5b area. Their priorities in regional economic policy were on the recovering of industrial competitiveness rather than employment. With the help of national and European aid, several manufacturing enterprises were modernized and newly created research and training centres succeeded in transferring new technologies and training workers in using them. No particular policy was devoted to the development of employment in the other sectors, such as retail trade, tourism or food industry despite their potential. The few initiatives taken to exploit the local resources and amenities are not yet visible. The highway to connect the region to the European road network is still not completed. Entrepreneurs tended to focus on the manufacturing industry, not the service sector. They did not develop specific strategies to maintain or augment employment and preferred flexible labour contracts. Labour unions of the region applied for INTERREG funds with their Belgian counterparts to finance a tourism project, but the regional authorities had not yet endorsed it at the time of the field survey.

Lessons and challenges

When the restructuring of its manufacturing sector began at the end of the 1970s, Luxembourg (B) had already a more favourable sector mix of agriculture, industry and services, was closer to a dynamic economic region lacking labour (the Grand Duché de Luxembourg) and had a more uniform spatial distribution of a growing population than Les Ardennes. In turn, Les Ardennes benefited from a better skilled manufacturing labour force with relatively lower costs than Luxembourg (B). What, however, made the difference in employment growth was the capacity of local actors in Luxembourg (B) to design and implement purposeful economic and employment strategies beginning in the late 1970s.

Thanks to a strong regional identity, a social and political consensus and an effective coordination, private and public local leaders in Luxembourg (B) were able to design and implement a territorial development policy. This concerted effort resulted first in strengthening the road network of the province and valorizing the comparative advantage of the province in terms of its central location, rural amenities and work ethic. In addition, this effort was successful in engineering attractive financial conditions for foreign investments and exploiting every financial opportunity provided by regional, national and European public funds. More recently, this concerted effort has been redirected to more directly supporting numerous small firms by providing equipped industrial sites, improving managerial skills specific to them as well as horizontal and vertical coordination between them and larger firms.

In contrast, the lack of political consensus among local leaders and the difficulty in building a common economic and employment strategy among the numerous municipalities in Les Ardennes diminished the comparative advantage embodied in a skilled manufacturing labour force with a relatively low cost. As a result, between 1980 and 1992, employment grew at an annual rate of 0.9% in Luxembourg (B) while it declined at an annual rate of 0.6% in Les Ardennes. Luxembourg (B) saw its unemployment rate decrease from 9.5% to 7.0% while Les Ardennes saw it increase from 13.9% to 14.4% between 1985 and 1995.

(1) Strengthening the capacity of local actors

The importance of strengthening the capacity of local actors to play an active role in developing and implementing together a common employment strategy is the primary lesson from this comparative analysis, a lesson also learnt from other cases examined in other parts of the European Union in this book. This capacity is reflected in the way in which local actors cooperate not only with each other but also with actors outside the region to design and implement specific projects in line with the strengths and opportunities of their region. In the case of Luxembourg (B), these specific projects included convincing national authorities to strengthen the road network, marketing the rural amenities of the region to potential investors and skilled managers from outside, exploiting every source of regional, national and European funds, building equipped industrial sites to facilitate the conversion or creation of firms and, more recently, providing risk capital and managerial training for managers of small firms.

The capacity of local actors can be strengthened by (Henry de Frahan *et al.*, 2000):

- providing local administrations with training and assistance in conceiving, planning, managing, monitoring and evaluating territorial development policy and assessing labour market development;
- re-organizing the work relationships between the different vertical and horizontal layers of local administrations to stimulate coordination and bottom-up initiatives;
- redistributing local and regional responsibilities among the different public and private institutions to avoid overlaps and implement the right development operation at the relevant level;
- organizing the private sector into a few trade unions and business associations to imply them in public-private dialogue and cooperation;
- and providing local entrepreneurs of small firms with training and assistance in management, market analysis, and technology assessment and financing.

(2) Strengthening internal and external networks

In line with this first lesson, strengthening internal as well as external networks is another key lesson. Internal networks in Luxembourg (B) were stronger than in Les Ardennes because of the active and responsible attitude of many local leaders, which was reinforced by a common sense of solidarity before specific employment challenges and a centralization of regional institutions. External networks between local actors and national and European authorities as well as foreign entrepreneurs and investors also performed better in Luxembourg (B) than in Les Ardennes as a result of strong internal networks and the relationships that local actors could build and use. Internal networks in Luxembourg (B) defined development priorities and organized practical aspects of firm settlements, *e.g.,* financial support, worker recruitment and training as well as social infrastructure for in-coming managers' families (education, health and entertainment). External networks were used to receive public funding and facilities and attract private investments and skilled labour. In sum, a strong capacity to design and implement sound employment projects associated with strong internal and external networks lead to employment growth in Luxembourg (B). Grouping municipalities, business associations, labour unions and other representatives of civil society into fewer institutions and organizing joint meetings on a regular basis facilitate the functioning of internal networks. By forming coalitions at the regional level, these representative groups can strengthen their bargaining power with regard to external public or private decision makers.

Both rural regions studied in this comparative analysis face several challenges. Since industrial restructuring, most employment growth in Luxembourg (B) has occurred in private services but, to a larger extent, in public services responding to population growth and in low skilled labour-intensive manufacturing activities dependent on subsidized investment. With public financial flows drying up and competition from low wage countries increasing, the sustainability of employment growth observed in Luxembourg (B) is at risk in both the private and public sector. Although employment in Les Ardennes relies less on public financial flows, it has increasingly relied on capital intensive manufacturing activities requiring less skilled labour. In the longer term, employment in Les Ardennes will depend on the extent to which managers of these high technology enterprises are able to maintain a competitive edge in their activities, while employment in Luxembourg (B) will rely on the extent to which small entrepreneurs are able to become modern managers of larger enterprises able to generate employment. The phasing out of the European structural funds and the reform of the Common Agricultural Policy as drafted in the Agenda 2000 in preparation for the integration of additional member countries from Central Europe will add stress to employment in both regions. The development of tourism, a positive avenue for both regions considering the increasing demand for rural amenities from urban regions, may, however, partly counterbalance this stress.

References

Barré, P. (1998) *Districts industriels (Industrial districts).* Mimeo, Institut des Sciences du Travail, Louvain-la-Neuve.
Benko, G., Dunford, M. and Heurley, J. (1997) Les districts industriels: vingt ans de recherche (The industrial districts: twenty years after). *Espaces et sociétés* no. 88/89, Editions L'Harmattan.
De Schrevel, M. (1994) Filière agro-alimentaire (Agrofood sector). *Entreprendre aujourd'hui, Supplément 7.*
Dupraz, P., Henry de Frahan, B. and Faucheux, G. (1998a) *Agriculture and employment in EU rural areas case study: the province of Luxembourg, Belgium.* INRA, Rennes and Catholic University, Unité d'Economie Rurale, Louvain la Neuve.
Dupraz, P. and Henry de Frahan, B. (1998b) *Agriculture and employment in EU rural areas case study: département Les Ardennes, France.* INRA, Rennes and Catholic University, Unité d'Economie Rurale, Louvain la Neuve.
Eurostat (1981-1997) *Regions; statistical yearbooks.* Luxembourg.
Eurostat (1985) *Statistiques de base de la Communauté (Statistics of the Community).* Luxembourg.
Eurostat (1996) *EUROSTAT Yearbook 1996; A statistical view on Europe 1985-1995.* Luxembourg.
Henry de Frahan, B., Dupraz, P. and Van Haeperen, B. (2000) Local capacity and industrial restructuring in the periphery of Belgium and France. In: Schaeffer, P. and Loveridge, S. (eds) *Small town and rural economic development: a case studies approach.* Greenwood, Praeger.
Institut National de la Statistique (INS) (1981) *Annuaires statistiques de la Belgique 1981 (Statistical handbook of Belgium 1981).* INS, Brussels.
INS (1991) *Annuaires statistiques de la Belgique 1991 (Statistical handbook of Belgium 1991).* INS. Brussels.
INS (1994) *Statistiques Régionales 1993 (Regional statistics 1993).* INS, Brussels.
INS (1996) *Statistiques Régionales 1995 (Regional statistics 1995).* INS, Brussels.

Institut National de la Statistique et des Etudes Economiques (INSEE) (1993) *Population, emploi, logement; evolutions 1975, 1982, 1990 (Population, employment, housing; evolution 1975, 1982, 1990)*. INSEE, Paris.

INSEE (1995) *Tableaux de l'économie Champenoise 1994/1995 (Review of the Champagne economy 1994/1995)*. INSEE, Direction régionale Champagne-Ardennes, Reims.

Organization for Economic Cooperation and Development (OECD) (1994) *Creating rural indicators for shaping territorial policy*. Paris.

German Case Studies: Niederbayern and Lüneburg

Heino von Meyer and Danielle Schrobiltgen

5

Selection of case study regions

Germany belongs to the most urbanized countries in the European Union (EU), together with the Netherlands, Belgium and the UK. Yet, since the unification of East with West Germany in 1989, about 20% of the German population live on 50% of the national territory, in local communities (*Kreise*) with population densities below 150 inhabitants per square kilometre. At a more aggregate level of territorial detail, the OECD has classified nine out of 47 German regions as 'predominantly rural', 18 as 'intermediate' and 20 as 'predominantly urbanized' (OECD, 1996). In 1996 the respective population shares were 8% (rural), 26% (intermediate) and 66% (urban). Due to the methodological approach of the RUREMPLO project, which aimed primarily at better understanding the dynamics of rural labour markets during the 1980s and early 1990s, this analysis will focus on a comparison of two rural regions in West Germany: Niederbayern and Lüneburg. Their performance in creating employment will be judged against the background of developments in other West German regions. Many findings, however, are also relevant for the rural regions in the New Länder of East Germany. The latter face today much bigger challenges of socio-economic restructuring and imbalanced labour markets. It is essential for them to learn from the positive as well as from the negative experiences in job creation and territorial development made in other German and European regions.

In this regard, a first important question arises: What defines an appropriate territorial context for rural development analysis and policy? Depending on the scale and characteristics of the territories considered 'rural', very different perceptions of problems and perspectives will be encountered, consequently leading to different policy conclusions and recommendations. For example: over the 1980s total population in Germany increased by 1.7%. If analysed at the level of local communities, 'urban' population, those living in communities with densities above 150 inhabitants per square kilometre, grew twice as much as population in rural communities (1.9% versus 1.0%). One could thus, conclude, that Germany faced a strong urbanization trend and a relative rural decline.

At the level of regions, however, the picture is not as clear. The 'predominantly rural' regions, those, with a majority of people living in low-density communities, had an equally high growth rate (2.2%) as the urbanized regions (2.1%), while intermediate regions faced stagnation (0.6%). This seemingly contradictory result can be explained as follows: The positive population growth of the predominantly rural regions was mainly due to a strong increase in their urban centres, small and medium sized towns (6.0%). In turn, in the urbanized regions, the rural communities surrounding urban agglomerations had particularly high growth rates (1.8%). This shows, rural development analyses and policies should focus neither exclusively on small individual rural localities nor on rural areas in general. The problems and prospects of rural communities need to be assessed in their wider regional context. Villages in remote rural regions have little in common with rural places in metropolitan regions facing urban sprawl. The important role of small and medium sized towns for the development of rural regions should not be ignored. Equally, the future development of many urbanized regions will only be sustainable if the problems and potentials of their rural communities are properly taken into account.

Given this background, comparatively large functional regions were selected for the German case study: Niederbayern and Lüneburg (Fig. 5.1). The region Niederbayern corresponds to the Bavarian administrative unit 'Regierungsbezirk Niederbayern', which is also a Eurostat region at level NUTS2. For Lüneburg it appeared inappropriate to analyse the entire Regierungsbezirk, since this is a very heterogeneous entity, where many local communities have functional links with neighbouring regions such as Hamburg, Bremen or Hannover. It was thus decided to compare Niederbayern to only the eastern part of the Regierungsbezirk Lüneburg: the Kreise (districts) Lüneburg, Lüchow-Dannenberg, Uelzen, Celle and Soltau-Fallingbostel. This region is smaller than the Eurostat NUTS2 region. For more than four decades, both regions, Niederbayern and Lüneburg, were situated next to the closed 'iron curtain', in comparatively peripheral locations – by German standards. Yet, they experienced very different development performances. While Niederbayern showed a very dynamic development, employment growth was only moderate in Lüneburg. Comparing these two rural regions promised to provide insights into some of the underlying causes of rural development success and failure.

Basic regional characteristics

Niederbayern is a comparatively large region, with over 10,000 km² and over 1 million inhabitants. Situated in the Southeast of Germany, bordering Austria and the Czech Republic, it consists of a less favoured, mountainous part in the East and a more fertile, hilly part west of the Danube valley. Nonetheless, as an administrative and functional region, it is perceived as an entity. Population is widely dispersed across the region. With 111 inhabitants per square kilometre population density is low, at least by West German standards (258 inh./km²). The three most important towns of Niederbayern (Landshut, Passau and Straubing, all

Figure 5.1   The German case study regions Niederbayern and Lüneburg

below 60,000 inhabitants) represent only about 10% of the regional population (Table 5.1).

Niederbayern has a rich variety of different cultural landscapes. With the Bavarian Forest, it hosts one of the few German National Parks. The region and its society have a profile, characterized by a distinct cultural identity and a strong sense of community. Remoteness has long been a major handicap, in particular for the eastern, mountain part of the region. Since the 1970s, however, transport infrastructure has improved significantly (motorways, canal, intercity train station).

Like Niederbayern, Lüneburg is a polycentric region, with a decentralized settlement structure. It has about 600,000 inhabitants and a density of only 85 inhabitants per square kilometre. The main urban settlements (Celle, Lüneburg and Uelzen) have between 35,000 and 75,000 inhabitants. Situated east of Hamburg, south of the river Elbe, Lüneburg was, until 1990, handicapped by its

Table 5.1 Population size of the main regional cities, 1995

Niederbayern		Lüneburg	
Landshut	59,100	Celle	73,500
Passau	50,300	Lüneburg	65,100
Straubing	41,700	Uelzen	37,200

Source: BBR, 1999.

location at the closed border between the two parts of Germany. This prevented new investments in the peripheral parts of the region. But even today, 10 years after German unification, the eastern parts of the region still lack proper access to major infrastructure networks. The major East-West transport links, Hamburg-Berlin and Hannover-Berlin, do not cross through and better connect the region.

Lüneburg is flat or hilly, with generally poor, sandy soils, a historical consequence of overgrazing and unsustainable forest use. The nature reserve 'Lüneburg Heath' and several other protected areas and nature parks are today important regional assets, not least for tourism development. Yet, the creation of a national park along the Elbe River had to be cancelled due to local opposition. Historical cities like Lüneburg or Celle preserve a rich cultural heritage. On the other hand, the region has suffered from the image of being a dumpsite for nuclear waste. Significant parts of the region are also military training area, thus not accessible to the public.

Both regions, Niederbayern and Lüneburg, have long been subject to regional/rural development policy support schemes, both from the Federal and the Länder level in Germany, as well as from the EU Structural Funds, in the context of 5b Programmes for rural development.

Some socio-economic features

Table 5.2 provides a first impression of basic characteristics of the two study regions. During the 1980s population growth in Niederbayern (8%) exceeded the West German rate (3%), while for Lüneburg the increase was below the national average (1%). Since 1990, with German unification and the opening of the eastern border, Lüneburg also experienced stronger population growth. Thus, by 1995, both case study regions had increased their population share in the national total, compared with 1980. Similar growth patterns are observed for employment and regional production (GDP). For Niederbayern GDP increased by a third over the 1980s and by half up to 1995. Over the same periods West German GDP grew by only 21% and 29%. For Lüneburg growth was lower, at 16% and 21%.

Table 5.2 Economic features of Niederbayern and Lüneburg, 1980-1995

	Niederbayern	Lüneburg	Germany[a]
Population, 1995 (million)	1.14	0.62	64.17
Area (1000 km²)	10.3	7.1	248.2
Population density (inh./ km²)	111	85	258
Population growth (1980=100)			
1980 - 1990	108	101	103
1980 - 1995	115	108	105
Employment growth (1980=100)			
1980 - 1990	117	102	107
1980 - 1995	124	107	107
GDP growth (1980=100)			
1980 - 1990	134	116	121
1980 - 1995	150	121	129
GDP per capita			
1995 (1000 ECU/inh.)	18.9	16.0	22.2
1980 (index)	78	77	100
1995 (index)	85	72	100
Unemployment rate (%)			
1980	4	4	4
1990	5	8	7
1995	8	12	11

Source: Own calculations based on Eurostat, BBR and NIW.
a) Old Länder of West Germany (excl. Berlin).

In 1980, in both regions, per capita income, measured as regional GDP per inhabitant, reached only about 77% of the West German average. Up to 1995 Niederbayern was able to improve its position to 85% of the reference level. Conversely, Lüneburg lost ground, with GDP per capita falling from 77 to 72%. Due to the low GDP levels in the New Länder, average per capita income for Germany as a whole, is today about 10% lower than that of the West German part. Thus, Niederbayern, only a few decades ago one of the poorest regions in Germany, has now almost reached the German average level of per capita GDP.

In 1980, regional unemployment rates for Niederbayern and Lüneburg were at 4%, equal to the West German national rate. Until the mid-1990s unemployment went up in all regions. For Niederbayern, however, a clear decoupling from the national trend can be observed: 8% compared with 11% in West Germany. The Lüneburg rate was even slightly above the West German level (12%). In the New Länder, however, unemployment rates are twice as high, in particular in the rural regions.

Employment change and sectoral mix, 1980-1995

As for rural economies in general, the sectoral employment mix in Niederbayern and Lüneburg is characterized by a comparatively high share of agriculture and forestry (Table 5.3). In 1980 they provided employment for over 20% of the workforce in Niederbayern, and for almost 15% in Lüneburg. By 1995 these shares had come down by 14 and 7 percentage points, to only 7% in both regions.

This is still more than twice as high as the national average. Yet, it can no longer be argued that agriculture would be the backbone of these rural economies. Today, also in rural regions nine out of ten persons employed work in industry or services. The future of the rural economy no longer depends primarily on agriculture. Dependencies are reversed: today, the viability of agriculture, the well-being of farmers and their families, relies heavily on a flourishing non-agricultural, rural economy. This is particularly true for regions such as Niederbayern, where a great majority of farmers are pluriactive, running their small farms only on a part-time basis, while having their main occupation outside agriculture.

Rural economies are today service economies. Over half of all employed persons in Niederbayern, in Lüneburg up to two-thirds, work in the service sector. In Niederbayern, 42% of the workforce and 30% in Lüneburg have employment in industry. The economic viability of rural regions depends no longer on milk or cereal prices, but on their capacity to offer opportunities for competitive manufacturing and innovative service delivery. Niederbayern and Lüneburg are clear examples. While at the national scale the employment share of manufacturing dropped by 6 percentage points, from 35% to 29%, both Niederbayern and Lüneburg were able to maintain their shares at around 34% and 22%. Thus, they increased their share in total national manufacturing.

At the national level, services in the financial and business sector showed the highest increase in shares between 1980 and 1995 (by 7 percentage points, from 22% to 29%). Both case study regions were able to keep pace with this expansion. Lüneburg's share is as high as the national average. For Niederbayern the share is slightly lower (25%), but the growth rate was higher (increase in share by 10 percentage points).

The fact, that rural regions become more and more industry and service oriented, does not mean, however, that their economies no longer show typical, specific features of rural economies. However, the characteristics that count for regional economic performance are no longer those that mattered in the past, when agriculture was the leading employer. Producing goods or providing services in peripheral, low-density regions, like Niederbayern or Lüneburg, is not the same as producing similar commodities, or offering identical services, in agglomerated metropolitan regions such as München or Hamburg. Rural economies still have their own features, their specific problems and potentials. They deserve particular attention.

Table 5.3 Employment growth, sectoral mix and regional dynamics in Niederbayern and Lüneburg, 1980-1995

	Niederbayern		Lüneburg	
	Sectoral shares	change[a]	Sectoral shares	change[a]
	1995 (%)	1980-95 (% points)	1995 (%)	1980-95 (% points)
Agriculture	7	-14	7	-7
Industry	42	-1	30	-3
Services	51	15	63	10
Mining, quarrying, electricity, gas, water	1	0	1	0
Manufacturing	34	1	22	-1
Construction	7	-2	7	-2
Trade, hotels, restaurants	13	3	15	1
Transport and communication	4	1	3	1
Finance and business services	25	10	29	8
Community and social services	9	1	15	2
Total employment (in 1000)	450	95	230	16
Change in total employment (%) due to:	+24		+7	
sectoral mix	-6		+4	
regional dynamics	+30		+3	

Source: Own calculations based on Eurostat, BBR and NIW.
a) Change in percentage points of share.

The need for a more careful analysis of the economic performance of rural regions can already be seen from the fact that regional development trends cannot automatically be derived from mechanistic models of economic growth and structural change. Niederbayern, for example, given its sectoral mix in 1980, with a particularly high share of persons employed in nationally declining sectors, such as agriculture or manufacturing, would have experienced a 6% decline in employment, up to 1995, if all branches had followed the national trends. In fact, however, Niederbayern was able to mobilize regional dynamics that more than overcompensated for the potential handicaps stemming from its sectoral mix. They actually resulted in a 24% employment growth. For Lüneburg, the job creation dynamic was less impressive (+7%). But also here the sectoral mix would have suggested an increase of only 4%.

Regional productivity and competitiveness

Employment growth cannot be the only relevant criterion for judging the development performance of rural economies. If regional production growth, the creation of value added, is not expanding at a similar, or even higher, rate, employment will not be secured in the medium to long term. To assess the competitiveness of regions, the growth of both employment and GDP, and thus implicitly also of labour productivity, should be analysed and compared with the national average rates of change.

In Fig. 5.2 information is provided for the two case study regions Niederbayern and Lüneburg as well as for the three types of regions: predominantly rural, intermediate and urbanized. On both axes the positive and negative differentials of regional to national growth are shown in percentage points of the standardized national growth indices. If a region is positioned in one of the two left/(right) quadrants, then their employment growth was lower/(higher) than the national average growth. In the two lower/(upper) quadrants GDP growth was lower/(higher) than national. On the diagonal cutting the square into two triangles, labour productivity growth is in line with national growth. In regions positioned in the upper left (lower right) triangle labour productivity has thus increased faster (slower) than for the national economy as a whole. Ideally a region is positioned in the darker shaded small triangle where production, employment and labour productivity all increased faster than national. This has actually been the case for Niederbayern over the period 1980-90. For 1980-95 Niederbayern is still showing above average growth for both GDP and employment, yet the dynamic of productivity growth became weaker, and from 1990 to 1995 was only average compared with national.

In turn, between 1980 and 1990, Lüneburg had lost position in both employment and GDP. Since German unification, the employment creation dynamic improved, up to 1995, yet at the expense of the labour productivity position that is now below average. The positive labour productivity changes during the 1980s in both Niederbayern and Lüneburg were primarily due to a shift from agriculture to more productive employment in manufacturing. The relative decline in the 1990s resulted from employment shifts to the less productive service sector and, in the case of Lüneburg, to a particularly strong boom in the construction sector. Growth in the construction sector was heavily induced by German unification, yet, there are severe doubts with regards to the long-term stability of the jobs created. Thus, it appears that the improvement of the relative position of Lüneburg since 1990, with regards to employment, is not based on solid ground and may quickly turn into another structural adjustment problem.

For an assessment of territorial development patterns in West Germany, in general, it is interesting to note that (the few) rural regions performed well with respect employment growth, although to some extent at the expense of labour productivity gains. Intermediate and urban regions stayed closer to the national average, with intermediate regions fairing well with regard to both employment and GDP, while urbanized areas lost both employment and GDP shares, and this negative trend even accelerated in the 1990s.

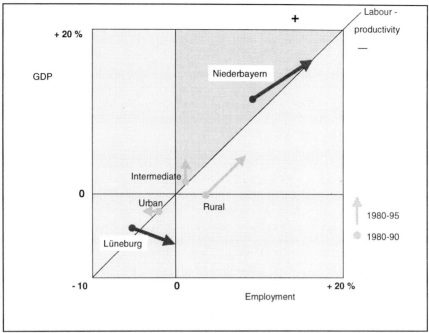

Figure 5.2   Regional competitiveness indicators: differentials to national growth in employment, GDP and productivity, 1980-1995 (%) [a]
a) Differences between the regional growth index and the national (West German) growth index in percentage points; national employment (GDP) growth: 1980-90: 7% (21%); 1980-95: 7% (29%).

Table 5.4 Regional development performance index: aggregate differentials to national growth in employment and GDP, 1980-1995 (%)

	1980-90	1980-95	Change 1990-95
Niederbayern	20	32	+12
Lüneburg	-9	-6	+3
Region types			
Rural regions	3	13	+10
Intermediate regions	3	3	0
Urban regions	-2	-3	-1

Source: Own calculations based on Eurostat, BBR and NIW.

If one assumes that growth differentials in GDP and employment were equally valued and can thus be traded-off (other assumptions could be made by assigning explicit weights), the sum of differentials provides a highly aggregate performance index for analyses of regional employment change. Like this, also changes over time can be assessed. Table 5.4 provides an example for such a comparison.

Human capital and innovation

Human capital and innovation are generally considered to be important explanatory factors for regional development performance. Education is often consequently seen as a priority for improving the employability of the workforce and thereby creating incentives for regional investment and job creation. Closer analysis shows, however, that these relationships are not as straightforward as it is often assumed. It is definitely not the average regional level of formal educational attainment, which counts for triggering dynamic regional/rural development. It is rather the quality of the skill match between regional labour supply and demand that matters. A rural region full of academics will not automatically face a bright future.

The Niederbayern case is particularly revealing in this context. Here the percentage of population with only primary education is one of the highest among all German regions. As Table 5.5 shows, only 14% of the population (aged 15 to 64) in Niederbayern have attained upper secondary education. This share is significantly lower than the national average (West Germany: 23%), and also clearly below the Lüneburg rate (18%). Also for tertiary education the percentage is low in Niederbayern.

Consequently the percentage of highly qualified employees (as classified by social security insurance) is as low as 2.6%, compared with 3.6% in Lüneburg and 5.6% on a national scale. Also this gap will not be closed in the near future, since in 1995 only 18% of the pupils in Niederbayern attended upper secondary schools; a percentage significantly lower than in Lüneburg (31%) where the share exceeded the national rate (23%).

Despite such statistics on formal education, entrepreneurs in Niederbayern do not normally complain about a lack of qualified workers or the human capital available in the region. On the contrary, the workforce in Niederbayern is considered to be well adapted to the needs of the local labour market. Thus, in Niederbayern the skill match between regional supply and demand profiles appears to be good. If and where needed to overcome bottlenecks, specific initiatives have been undertaken to offer and promote targeted training on the job. Here the regional chamber of commerce, the labour office and the trade unions have been particularly active in setting-up training schemes, etc. The quality of the regional knowledge base does not depend solely on individual learning and formal educational attainment. Vocational training and informal learning are of equal importance for ensuring a proper match of skills.

Beyond individual learning, collective, organizational and institutional learning become increasingly relevant. Innovation in firms and public

Table 5.5 Education and innovation in Niederbayern and Lüneburg

	Niederbayern	Lüneburg	Germany[a]
Educational attainment of inhabitants aged 15-64, 1990 (%)			
Upper secondary	14	18	23
Tertiary	6	7	9
Pupils, 1995 (total=100)			
Upper secondary	18	31	26
Tertiary	7	8	18
Students, 1992 (per 1000 inh.)	11	9	27
Highly qualified workforce, 1989 (%)	3	4	6
Patents per million inhabitants, 1990-1992	100	78	217
Technology intensity of industry (%)			
High and Medium-high	51	44	50
Medium-low	22	18	28
Low	28	38	22
Innovation type of industry (%)			
Supplier dominated	37	39	36
Scale intensive	42	38	32
Specialized plus science based	22	22	33

Source: Own calculations based on Eurostat and BBR.
a) Old Länder of West Germany (excl. Berlin).

administration as well as the maintenance and improvement of an attractive entrepreneurial climate are decisive for stimulating regional rural development approaches. A simple indicator for the innovative capacity of a regional economy can be seen in the number of patent applications per inhabitant. Here Niederbayern scores better than one could have expected from the educational statistics. Obviously rural regions like Niederbayern or Lüneburg cannot reach the average national intensity (220 patents per million inhabitants), yet with an average of about 100 patents Niederbayern performs better than Lüneburg with less than 80.

It is also interesting to note that the technological profile of the industry sector in Niederbayern is not at all characterized by an overrepresentation of low-tech branches. On the contrary, in 1990 51% of the workforce were employed in branches classified as high-tech or medium high-tech. The share of low-tech employment was 28%, compared with 38% in Lüneburg, and 22% for the West-German average. While this branch classification is focused primarily on aspects related to product innovation, a more process oriented classification shows that both Niederbayern and Lüneburg have rather low employment shares in specialized, and science-based branches (22% as compared with 33% nationally).

This analysis reconfirms important findings for the promotion of regional/rural development: neither is a high skill regional workforce an automatic guarantee for a high growth, high-tech economy, nor do high-tech branches automatically provide or require high skill jobs. They might even look for low skill, low pay employment. Yet, any simplistic policy conclusions should be avoided. If during the 1980s and early 1990s the qualification levels of the rural workforce in Niederbayern and other rural regions were adequate, this will not necessarily continue to be the case in the future. The phase when manufacturing employment shifted from urban to rural regions may well have come to an end. Today, the competition from less developed regions inside and outside the EU is felt more and more strongly. In the future rural manufacturing in Germany may only be sustainable, if it is able to compete as a result of higher product or process quality rather than on the basis of lower prices and costs.

Social capital and regional identity

Regional innovation and development processes rely not only on individuals alone, but rather on the quality of their interaction shaped by regional traditions, history, culture, etc. Consequently development analysis cannot focus on human capital alone, but has to look beyond, trying to capture aspects of social capital such as trust, cohesion, willingness and capacity to cooperate.

A crucial factor for the development success of Niederbayern has been its strong regional identity, a common perception shared by a great majority of regional actors in business as well as public administration. With respect to economic development, this common view is probably more rooted in a consciousness of (past) regional disadvantage than in a shared vision of the future. Yet, historic disadvantages were, to some extent, turned into regional assets, capital for future development. For example, the most remote parts of the region now host a national park that attracts a great number of tourists. They find accommodation in rural places and landscapes, which have for decades been shaped by pluriactive farm families. These populations formed a highly reliable workforce with a strong sense of responsibility, which made it attractive for entrepreneurs from both inside and outside the region to invest there.

The strong regional identity and image of the Niederbayern region finds its expression also in the territorial organization of public administration as well as of non-governmental institutions. Regional actors operate at the same scale for the same region. They were, thus able to deliver a strong coherent message in lobbying for investment and support from outside.

For the Lüneburg region this is not the case. The relevant regional administration, the *Bezirksregierung*, has to cover a much bigger and highly heterogeneous area (the NUTS2 region). While some parts of the region are influenced by Hamburg, others are oriented towards Bremen or Hannover. The less developed rural areas in the western part of the *Regierungsbezirk* have little in common with the case study region in the East. They may share problems, yet, they are too distant to form a meaningful entity for regional economic development. Not only does the public administration not match with the regional division, also the labour office and non-governmental institutions such as the chamber of commerce deal with territories, which, in different directions, reach beyond the case study region.

Given this lack of an adequate regional entity, the role of the Districts (*Kreise*), as the next (lower) layer of public administration, is rather important. However, since the *Kreise* tend to compete rather than cooperate, in particular with respect to economic development and attracting investment, this implies an even weaker regional image. The lack of a coherent vision for the Lüneburg region must be considered as a key reason for lagging behind in development.

Regional rural policy

The development success of Niederbayern can be attributed in part to its endogenous development potentials, in particular to a strong regional identity combined with an effective institutional structure and networks; the endogenous development engine needed exogenous fuel, however. Without substantial, long-term support from other layers of government, from the *Land* Bavaria, from the Federal and even from the European level, the local/regional administrations and actors would not have had the resources for a similar take-off.

Niederbayern and Lüneburg have long been covered, and partly still are, by the national regional policy scheme (GRW, Joint Task) co-financed jointly by Bund and Länder. This was sometimes also supported by contributions from the EU Structural Funds, in particular under Objective 5b (rural development). The success of Niederbayern cannot be attributed, however, to regional policy alone. Much more broadly designed strategies and measures for territorial development were put in place.

In Niederbayern the economic development strategy clearly followed a territorial, regional approach. Indicative regional (development) plans, launched and coordinated at the level of the *Land* Bayern, gave guidance to public and private investors and served as a basis for coordination and cooperation. In the 1970s strategic infrastructure decisions were realized concerning the construction of highways, universities, spas, etc. Contrary to the growth pole, export-base, and specialization concepts of the regional policy promoted by the Federal level, Bavaria did not concentrate investment aid exclusively on regional centres. Early emphasis was put on the support of small and medium sized enterprises (SMEs), on diversification and technology transfer. This did not exclude, however, massive support provided also to large-scale projects (BMW).

Another strategic decision in the early 1970s has been to set up universities and higher technical schools in the peripheral rural parts of Bayern. Initially, this had no immediate impact on the development of local firms in the region. Meanwhile, however, it is acknowledged that these academic institutions have become important centres fostering economic development and technology transfer in the region. In agricultural policy too, Bavaria went its own way, refusing the priorities of Federal and EU agricultural structures policies, which aimed at support for large scale, economically viable full-time farms. Support was always provided also to pluriactive farm families that represent the majority of the farming population in Niederbayern.

Lüneburg too always formed part of the main regional policy schemes operated in Germany. Thus, it is not access to regional policy as such that can explain the difference in rural development performance. It is rather the use regional actors were able to make of the policy options offered. Beyond the already mentioned lack of cohesion among regional actors, another important difference is relevant in this context. While Niederbayern benefited from a delocalization process out of the München agglomeration, which was actively accompanied by an explicit Bavarian policy, Lüneburg did not benefit from a similar outflow of investment from Hamburg. Hamburg, being a *Bundesland* on its own, has no interest in promoting decentralization of economic activities into the neighbouring *Bundesländer* of Schleswig-Holstein and Niedersachsen. On the contrary, Hamburg undertakes a lot to keep investment within its urban perimeter. The *Land* Niedersachsen, to which Lüneburg belongs, does not have a comparable urban agglomeration with similar 'trickle-down' effects, nor has it in the past established a territorial development policy as explicit and elaborate as Bavaria did.

Concluding remarks

From the analysis of employment creation dynamics in Niederbayern and Lüneburg, two of the most rural regions in West Germany, a few general findings can be derived:

(1) Over the 1980s and the early 1990s some rural regions in West Germany have shown exceptionally good performance in employment creation and GDP growth. Niederbayern, only a few decades ago a poor, peripheral rural region lagging behind, has meanwhile become a dynamic, prosperous regional economy. It is an encouraging example for other rural regions in Germany and Europe, although lessons cannot simply be transferred. A better understanding of the underlying regional strengths and dynamics may help to improve the effectiveness and efficiency of rural development strategies and policies designed and implemented by the various local, regional, national and supranational actors.

(2) The case studies prove that the chances for positive development performance are better if the territorial perimeter of the relevant public and private institutions involved in the regional/rural development process, be they public administrations, labour offices, chambers of commerce, or other non-governmental organizations, match the same territory. This helps reinforcing regional identity, creating a commonly shared development vision, cooperating and coordinating joint development efforts. Where such a territorial match is achieved, it seems easier to set up networks of partners, strengthening coherence and cohesion internally and mobilizing external support from outside the region.

(3) Agriculture is no longer the driving force of economic development in rural regions. Its employment share is below 10%, its contribution to the regional value added is below 5%. Both shares are constantly declining. Yet, together

with forestry, agriculture is still managing about 80% of the land area, thus shaping the visual amenities and the ecological qualities of the diverse cultural landscapes. This contribution to rural development is increasingly valued, but not automatically paid for. If properly managed and promoted, amenities can become major rural development assets (national or regional parks, scenic landscapes, cultural heritage, etc.). However, it is not so much the existence of rural amenities that matters, but more the way they are used to trigger rural employment and value added, be it through tourism, through marketing of local products, or as a locational value for attracting new populations and investors.

(4) Rural economies have become service economies. The majority of jobs and an even greater share of the new employment opportunities originate from the service sector. Yet, this is true for all regions. The main difference explaining the above average performance of Niederbayern, as compared to Lüneburg is an exceptionally good performance in maintaining and attracting industrial employment. It is unlikely that in the future similar potentials of manufacturing investment can be mobilized in favour of the rural areas in Germany. Rural employment policies should, however, encourage careful analyses of the respective strengths and weaknesses across the entire employment spectrum instead of focusing too narrowly on traditional rural sectors alone.

(5) Overall, the analyses have shown that the fate of rural economies is not determined by their sectoral mix. Sectors, which, in aggregate, are declining nationally and internationally, may well be drivers of positive employment growth in some regions. Since the sectoral mix explains only a small part of total change in rural employment, more emphasis should be put on better understanding and fostering those factors that are responsible for region specific development dynamics.

(6) Human capital is one factor that could potentially explain rural development differentials. This cannot be measured, however, in terms of formal educational attainment only. Niederbayern performed particularly well despite the fact that it has a comparatively low share of highly qualified labour. The profile of industry in Niederbayern is characterized by an above average share of firms in high-tech and medium high-tech branches. High-tech firms do not necessarily offer high skill jobs. What matters is not the qualification level of labour supply alone, it is rather the match of skills offered and demanded within the region. Past experiences should not simply be transposed into recommendations for the future. However, rural development initiatives, which aim at improving the human capital base, should carefully assess the bottlenecks and priorities on the labour demand side too. Strategies should not focus exclusively on formal schooling but should address a wider spectrum of skills, encourage cooperative training initiatives and promote advanced learning by doing.

(7) Although the case studies show, that differential performance in rural development in general, and employment creation in particular, tends to be driven by specific regional dynamics that rely heavily on endogenous potentials, this should not lead to problematic policy conclusions. For decades, both Niederbayern and Lüneburg have benefited from various regional and rural policy measures, as well as from strategic territorial development decisions concerning infrastructure development etc. Those were taken and funded by various levels of administration (local, regional, Länder, Federal and European). Comparisons between two regions can thus only 'explain' the differential on top of a ground layer of support, which surely had an impact on the basic development trends. This analysis has not looked at a 'policy-off' scenario. Consequently, the findings of this comparison should not be misinterpreted as a plea for abolishing basic rural development support. On the contrary, consideration should be given to if and how the existing support measures could be reshaped in a way that would encourage the generation and clever use of regional development potentials more effectively.

References

BBR (Bundesanstalt für Bauwesen und Raumordnung and BfLR Bundesforschungsanstalt für Landeskunde und Raumordnung) (1992-1999) *Laufende Raumbeobachtung; Aktuelle Daten zur Entwicklung der Städte, Kreise und Gemeinden (Ongoing space monitoring; Statistics on the development of cities, districts and communities),* various issues. Bonn.

Bollmann, R.D. and Bryden, J.M. (1997) *Rural employment; An international perspective.* CAB International, Wallingford, UK.

Eurostat (Statistical Office of the European Communities) (1985-1997) *Regions; statistical yearbook,* various issues. Luxembourg.

Irmen, E. and Blach, A. (1996) Typen ländlicher Entwicklung in Deutschland und Europa (Types of rural development in Germany and Europe). *Informationen zur Raumentwicklung,* 1996-11/12, 713-728.

NIW (Niedersächsisches Institut für Wirtschaftsforschung e.V.) (1997) *Regionalbericht 1995/96/97 (Regional report 1995/96/97).* Hannover.

OECD (Organization for Economic Cooperation and Development) (1996) *Territorial indicators of employment; Focusing on rural development.* Paris.

Von Meyer, H. (ed.) (2000) *The dynamics of rural employment creation; Case study leading region Niederbayern and lagging region Lüneburg.* PRO RURAL Europe, Wentorf (Hamburg).

Employment Dynamics in Rural Greece: the Case Study Regions of Korinthia and Fthiotis

6

Sophia Efstratoglou and Angelos Efstratoglou

Introduction

'To increase employment in the European Union is still the first fundamental challenge facing Member States...' (European Commission, 1996). The need to increase employment-creating growth is one of the priority objectives of the European Employment Strategy, with the intention that the achieved output growth is translated into more jobs in both urban and rural regions of Europe. Within these priorities, the identification and enlightenment of the driving forces underlying employment increases and job creation in rural areas has become an important issue on the European rural development policy agenda.

This issue takes an even higher priority for Greek rural areas, which maintain significant human resources and still have a high dependence on agriculture. About 40% of the country's population lives in rural areas[1] and more than one-third of their labour force is in agriculture (NSSG, 1995). Modernization processes in agriculture and pressures on farm incomes, due to CAP reforms, have accelerated the exodus from farming. Some rural areas have succeeded in absorbing the exodus from agriculture through expansion of their non-agricultural labour markets and have increased their total employment, while other rural regions' productive systems have not shown the capacity to generate new job opportunities and to absorb the exodus from agriculture. In such cases, the agricultural exodus often has been followed by depopulation, which threatened even more their fragile socio-economic fabric in rural areas.

Within this framework, the objective of this chapter is to present the key findings with regard to employment increases or stagnation and the determining factors and driving forces behind these employment changes in the two Greek case study regions of the RUREMPLO project (Efstratoglou *et al.*, 1998a, b). Moreover, understanding the in-depth employment dynamics in a leading rural

---

[1] Rural areas in Greece are defined as the areas where population resides in communes of less than 10,000 inhabitants (NSSG, 1996).

region allows very often for the possibilities of transferring successful performance examples to lagging rural regions.

Selection of case studies

Greece consists of 13 NUTS2 regions, of which ten are classified as most rural, one as intermediate and only two as most urban (Chapter 3). Five of these regions had a high non-agricultural employment growth (measured as deviation from the national average) during the period 1981-1991, and only two experienced an employment growth below the national level. However, considering the significant variation and heterogeneity in geography, land morphology and local resource structures, that prevail within a NUTS2 region, it was decided for Greece, that NUTS3 level regions are more appropriate for the case study analysis. Greek NUTS3 regions, which are 51 in total, are territorial and administrative units (prefectures). In the case of Greece, except for the criterion of relative change in their non-agricultural employment, the selection of the case study regions was also based on the capacity of the region to absorb the large exodus from agriculture. Thus, in addition to the performance in non-agricultural employment, the increase/decrease in total employment was also taken into consideration. Based on these two criteria, Korinthia was chosen as the leading case study region and Fthiotis as the lagging one (Table 6.1). It should be noted that the selection of these two case studies, besides the above mentioned criteria, challenged our knowledge as both regions are rather similar with regard to their rural-urban structures, sectoral employment and to their location as non-remote regions.

During 1981-1991, Korinthia, as a leading case study region, experienced a significant increase in its non-agricultural employment (3% annually) compared with only 2% in Fthiotis. At the same time, Korinthia's productive system succeeded in absorbing the agricultural exodus and increased its total employment above the national average. On the other hand, Fthiotis did not manage to absorb the agricultural exodus and also due to its industrial employment decline, its total employment decreased by 0.6% annually.

Table 6.1 Employment and population growth, 1981-1991 (% p.a.)

	Korinthia	Fthiotis	Greece
Total employment	0.8	-0.6	0.5
Agricultural employment	-2.5	-4.0	-2.5
Industrial employment	0.0	-1.5	-1.3
Services employment	5.1	4.0	3.7
Non-agricultural employment	3.1	1.9	1.8
Population	1.4	0.6	0.5

Source: National Statistical Service of Greece, Population Census, 1981, 1991.

Profiles of Korinthia and Fthiotis

Korinthia is located in the northeastern part of the region of Peloponissos at a distance of 85-175 km from Athens, while Fthiotis is in the region of Sterea Ellada, at a greater distance of approximately 180-300 km from Athens (Fig. 6.1). Korinthia is smaller in size than Fthiotis (Table 6.2), it has less population but a higher population density, as its rural areas are more densely populated.

Korinthia can be characterized as a more dynamic area compared with Fthiotis in many respects. With regard to demographic changes, it showed a significant increase in its population during 1981-1991, almost three times that of Fthiotis and the national average. This was mainly due to a higher natural population growth and immigration. This dynamic trend was also recorded for Korinthia's non-agricultural and total employment increase. During the same period, real GDP growth was also at a higher level in Korinthia (2%) compared with only 1% for Fthiotis. This development pattern contributed to higher levels of per capita income in Korinthia, compared with Fthiotis and to the national level, for the whole period of analysis (Table 6.2).

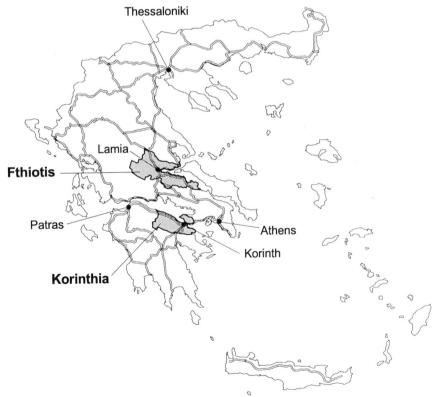

Figure 6.1   Location of Korinthia and Fthiotis in the Greek context

Table 6.2 The economic profile of Korinthia and Fthiotis

	Korinthia	Fthiotis	Greece
Population, 1991	141,800	171,300	10,259,900
Distribution of population (%)			
Level areas	59	53	69
Hilly areas	20	34	22
Mountainous areas	21	13	9
Size (km²)	2,290	4,440	131,960
Population density (inh./km²)	62	39	78
Real GDP growth, 1981-1991 (% p.a.)	2.2	1.0	1.9
GDP per capita (ECUs)			
1981	4,000	3,300	3,100
1991	5,400	4,200	4,800
Unemployment rate (%)			
1981	3	3	4
1991	7	9	8
Population main towns, 1991			
Korinthos:	27,400	Lamia: 44,100	
Loutraki:	9,400	Atalandi: 6,200	
Kiato:	9,100	Stylida: 5,000	

Source: National Statistical Service of Greece, Population Census 1981, 1991 and National Regional Accounts of Greece.

The rural/urban structure of both rural regions is typical for that of most Greek rural regions, which is characterized by the presence of a medium size town (in Greek standards) that concentrates 20-25% of the region's population. In Korinthia, the main urban centre is Korinthos with 27,400 population (19% of total population); in Fthiotis, it is Lamia with 44,100 population, (25% of total population). In both regions, there are also two or three small towns with a population less than 10,000 inhabitants. In general, Korinthia's overall level of development is much higher than that of Fthiotis. According to its 'development index', the region is ranked on the 13th place among the 51 prefectures of the country, while Fthiotis is ranked on the 26th place (Athanasiou *et al.*, 1995).

Local resources

Both case study regions are endowed with significant natural and tourism resources, but there are considerable differences in the degree of their exploitation. Korinthia has exploited its local resources (natural, agricultural, tourism) to a large degree, while this has happened too a much lower degree in Fthiotis. Korinthia has also a better road infrastructure, which combined with its approximation to Athens, creates a significant comparative advantage for the region. More analytically, the contribution of local resources and infrastructure to employment creation in the case study regions, appears to be as follows:

In Korinthia, local resources have substantially contributed to employment creation and to its development process. During the period of analysis (1981-1991), resource-based activities increased their employment (with the exception of agriculture, which however recorded a slower exodus compared with Fthiotis and the national average). Korinthia possesses important natural resources like fertile land, forests of recreational value and beautiful landscapes in both mountainous and level areas. Agrifood industries have developed activities in the field of wine, olive oil, canned fruits, raisins, cheese and meat processing and spring water bottling. Tourism is an important economic activity for the region. Main tourist resources are the lengthy sandy beaches along its 80-km coastal line, its mountain areas covered with forests, its natural biodiversity, villages with traditional architecture and archaeological sites. Mountain areas have potential for winter tourism. The tourist capacity in the region amounts to 8,200 beds and tourist overnights spent were 651,000 in 1994. The tourist dynamism of the region is reflected in the 22% increase in hotel beds and the 43% increase in tourist overnights between 1981-1991. Due to its proximity to Athens, the region has also attracted population from Athens for a second residence. The road infrastructure in Korinthia is better developed compared with other rural regions in Greece. The national highway (Athens-Thessaloniki-Patras) expands along the region. Two other main regional routes connecting Athens to the southern part of Peloponissos region cross Korinthia as well. However, road infrastructure in mountainous areas is inadequate. Infrastructure in the three small ports and the Korinth Canal facilitate the region's trade of oil refineries and the exports of its agricultural products.

Fthiotis has also important natural, mineral and tourist resources. Economic activities related to natural resources generated significant employment in the past. Nowadays, they continue to employ a significant share of the labour force in farming, agricultural processing industries and textiles linked to the local cotton production and in mineral industries. However, during the study period, all resource-related economic activities had a decline in their employment. Natural resources refer mainly to fertile land, forests of recreational value, beautiful landscape, particularly in the mountainous areas, and an unspoiled environment. The region has also mineral deposits of ferro-nickel, chromium and bauxite around which metal industries are concentrated. Tourism resources consist mainly of seashores, therapeutic springs for spa tourism, historical sites, natural protected parks and traditional mountainous villages, but these remain to a large degree non-exploited. Despite the availability and potential of tourist resources in Fthiotis, the number of tourist overnights in the region declined by 32% during 1981-1991, mainly due to a lack of tourism strategies such as up-grading hotel standards, exploiting the spa tourism potential, advertising the region or promoting its local identity. The region has an average infrastructure compared with other rural regions in Greece. However, the national highway axis (Athens-Thessaloniki), that crosses the region, has contributed to encouraging the settlement of new industrial enterprises. The region's location – in the middle of the country and almost equidistant from the two major cities of Athens and Thessaloniki – is another factor that has contributed to attracting exogenous economic activities.

Table 6.3 Sectoral distribution of employment, 1981-1991 (%)

	Agriculture		Industry		Services	
	1981	1991	1981	1991	1981	1991
Korinthia	47	33	24	22	29	45
Fthiotis	48	34	22	20	30	46
Most rural regions	48	34	23	22	29	44

Source: Efstratoglou *et al.*, 1995.

Economic activities

Economic structures and organization are quite similar in both case study regions. The sectoral distribution of employment is almost the same (Table 6.3) and both regions maintain still significant human resources in agriculture (one-third of their labour force).

Other similarities concern the structure of enterprises and the resulting social formation of employment. More analytically, the small size of enterprises, which is the main feature of the Greek economy, prevails in both regions. In both regions, about 97% of enterprises employ fewer than 10 persons. During the period of analysis, new jobs created in both case study regions were mostly in new, mainly small enterprises, while employment in large and medium enterprises declined, mainly due to technological adjustments. The structural characteristics in both regions have contributed to a large number of self-employed (40% of total employed in 1991 in Korinthia and 42% in Fthiotis) and low levels of salaried and waged employees (42% and 45% respectively). At the same time, the relatively higher number of employers (over 7%) in Korinthia compared with nearly 5% in Fthiotis, indicates the existence of more local entrepreneurial opportunities in Korinthia.

Despite the common structures, the dynamics of employment by sectors and branches of economic activity during the period 1981-1991 were significantly different (Fig. 6.2). Korinthia had a slower exodus from agriculture (-22%) compared with Fthiotis, which lost about one-third of its labour force in agriculture. For the whole country, the agricultural exodus amounted to 29% in this period. The slower exodus from agriculture in Korinthia can be attributed to its more labour intensive production patterns and higher farm incomes, which tend to maintain better living conditions for farmers. Fthiotis' agricultural structures are in general more extensive with lower value added per employed (62% of that of Korinthia). At the same time, Korinthia managed to maintain its employment in the secondary sector, despite the industrial recession period that the country was going through. Fthiotis was caught up in this recession process and experienced a significant decline in its secondary sector employment (-14% compared with -12% for the country). This was mainly due to the lower

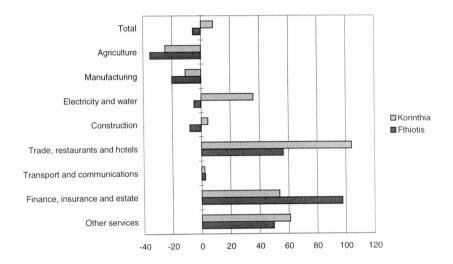

Figure 6.2   Employment changes by branch in Korinthia and Fthiotis, 1981-
1991
Source: Derived from Population Census Data 1981,1991.

competitiveness of the mix of branches in its industrial sector and to the
restructuring process that started during this period.

The service sector in both regions had the leading role in job creation, due to
the tertiarization of their economies. However, in Korinthia the service sector
showed a higher dynamism with an employment increase of 65% in the period
1981-1991, while in Fthiotis, it was somewhat less (48%).

From the above it becomes obvious that there are strong restructuring
processes in both regions, which reflect competitive forces that underlie local
economies as well as the capacity of local actors to adjust to the new
environment. In this restructuring process, there are economic branches that are
gaining jobs and branches that are losing jobs.

In Korinthia, branches that showed considerable employment increases were
trade-hotels-restaurants (104%), finance-insurance (53%) and other services
(61%), which concern mainly entertainment, education, health, public
administration (Fig. 6.2). The demographic dynamism of the region, the increase
in tourism, the higher farm incomes and the maintenance of jobs in the secondary
sector could be considered the main forces behind this overall service sector
growth. On the other hand, branches that lost jobs in Korinthia were agriculture
(-22%) and manufacturing (-10%). It is useful to note that the decline in
manufacturing employment in Korinthia was quite low compared with that of
Fthiotis (-20%). Main losing branches in manufacturing were electric machines,
machinery-equipment, textiles, paper and plastic products. However, the
contribution of those branches in Korinthia's total labour force was low.

In Fthiotis, branches with high rates of employment growth were also banking-insurance (98%), trade-hotels-restaurants (57%) and other services (50%). Branches that lost jobs were agriculture (-35%) manufacturing (-20%) and mining (-40%). It is important to note that manufacturing employment in Fthiotis declined as a result of a de-industrialization trend in Greece during the 1980s. This trend appears to have continued in Fthiotis in the 1990s. As the industrial mix mainly consisted of branches more exposed to de-industrialization trends, this seems to be the main factor behind employment decline. Textiles, clothing and leather, non-metallic mineral products and machinery equipment, which had more than half of the total employment in manufacture of the region in 1981, were mainly affected by de-industrialization trends. They showed significantly high rates of employment decline (-48%, -40%, -45% and -76% respectively) in the 1980s, while the same branches in Korinthia, constituted only 25% of employment in manufacture (in 1981) and moreover showed lower rates of employment decline. The evolution in manufacturing has very often tended to be considered as 'change initiating' rather than 'change following' (Frost and Spence, 1991), in the sense that its evolution causes changes in the other sectors of the economy and especially in services. So, the large and continuing decline in manufacturing employment in Fthiotis could be considered as a key factor behind lower growth in services and total employment decline.

Labour supply

Population increase in Korinthia in the 1980s was significantly higher (15%) than that of Fthiotis (5%) and this affected changes in labour supply. Korinthia's demographic dynamism, mainly due to immigration, may be considered as a key factor behind the increase in its labour force (20%). However, the participation rate of the population (15-64 years old), which was a little higher in Korinthia (56%) than in Fthiotis (53%) in 1991, showed a common declining pattern in the 1980s in both regions. This was mainly caused by the decline of the participation rates in the very young (below 20 years), due to the increasing entrance of young population into education, and the expansion of the social security system for the older age group (above 55 years). In the same period, participation rates of population (15-64 years old) in Greece increased from 54% to 55%, due to the increase in the rate of women from 29% to 35%, while the rate of men declined from 79% to 75%. On the contrary, the increase in labour force in Fthiotis was only 5% (at similar levels with population change).

Large differences in the participation rates by gender (79% for men and 32% for women in Korinthia and 76% for men and 30% for women in Fthiotis) may be considered as a main feature in both regions. It is useful to note that although participation rates for women in the country showed a significant rising trend in the 1980s, both regions did not manage to benefit from this, as Korinthia showed a slight rising trend, from 29% to 32%, while it declined from 31% to 30% in Fthiotis. This indicates the lack of employment opportunities for women in rural regions, as well as the existence of certain social patterns of division of labour in the family and more generally in society (Overbeek *et al.*, 1998).

The educational level in both regions is increasing, but is still at a low level. This is mainly attributed to the significant presence of agriculture. Korinthia seems to have a slightly better educated labour force, but this does not seem to influence employment changes, since the availability of skills in the labour force is of most importance. With regard to the local labour force skills, they seem to meet the needs of enterprises, although in general there is room for further improvements in both regions. Of course, hiring labour from local labour markets is not related only to the availability of the necessary skills, but also to employers' preference of hiring local people for reasons concerning workers responsibility, absenteeism, etc. Moreover, as almost all enterprises interviewed pointed out that they prefer on-the-job training procedures, the lack of skills does not seem to be a factor that hampers employment growth. In other words, any mismatches in demanded and supplied skills do not seem to constrain employment growth.

Both regions show almost identical unemployment patterns. Unemployment rates rose from 3% in 1981 to 7% in 1991 in Korinthia and from 3% to 8% in Fthiotis in the same period. The rise in women's rate was sharper, from 5% it increased to 13% in Korinthia and from 4% to 14% in Fthiotis. For men unemployment rates were much lower and started from almost 3% to 5% in Korinthia and from almost 3% to 6% in Fthiotis. Data from administrative sources (registered unemployment) provide evidence that in both regions unemployment continues to rise in 1990s for men and women as well. Almost three-quarters of the unemployed in both regions are under 30 years old, providing evidence of the difficulties that young people face entering the labour markets and of the structure of employment in rural regions. However, this almost identical pattern of unemployment might be misleading, as someone may think, that both regions are showing similar difficulties and similar lack of employment opportunities. Nevertheless, it should be mentioned that unemployment rates are not always the most appropriate indicators of economic performance (employment opportunities) in a region. This is because low unemployment rates may be accompanied by a 'workers discouraging effect', an effect that in periods of economic crisis and lack of employment opportunities drives workers out of the labour force (Beatty and Fothergill, 1996). In general, it can be concluded that although both regions show a large number of similarities, the deterioration of employment evolution in Fthiotis is not revealed through unemployment rates, but rather through lower participation rates and hidden unemployment.

Labour markets

The formation, structure and functioning of the labour markets in both regions appear to have common features, due to the existence of common explaining factors. The significant presence of self-employed and the large numbers of small family-based enterprises prevent the labour market from being the main mechanism for the allocation of the labour force among economic activities. Moreover, the highly segmented structure of the labour markets that arises mainly from prevailing economic structures, and the social formation of production affect the ways in which employment growth is generated. These highly segmented structures become increasingly important not only for the number of jobs created, but also for their type and characteristics.

Considering the high presence of self-employed (farmers, professionals and small family business), the labour market functions concern about 40% of the total labour force in both regions. At the same time, their labour markets are characterized by a great degree of flexibility, despite the existence of regulations (Institute of Labour, 1999). Moreover, the prevailing small size of enterprises and personal contacts contribute to the non-compliance with labour market regulations. It should be noted that labour regulations apply only to the private sector, as in the public sector labour relations are covered by regulations of public law (European Commission, 1997).

As far as wages are concerned, the average remuneration in manufacturing is about 10% higher in Korinthia than in Fthiotis. Comparing the wage levels of different manufacturing branches with changes in the work force in Korinthia, it can be seen that branches that experienced the most significant decline in employment, have in some cases lower, and in other cases higher wages than the average level. Thus, a direct relation between wage levels and employment changes is not observed in Korinthia. On the other hand, in Fthiotis it is observed that all branches with a significant decline in employment have lower wages than the region's average, while branches with employment increases have higher wages. Moreover, despite its lower average wages in manufacturing, Fthiotis is losing relatively more jobs than Korinthia. Based on that, it can be argued that labour cost is not the main factor behind employment decline. In general, employment growth does not seem to be hampered by institutional rigidities, since labour market functioning is quite flexible, not only with regard to regulations, but with regard to wage levels as well.

Actors

Although both regions have a common policy framework in the scope of their status as Objective 1 regions, the capacity of the local actors involved in the implementation of these policies seems to be different. The same applies to their capacity to develop strategies as well as to their ability to exploit opportunities through the interplay of internal and external networks. Main features on policies, strategies and networks in the regions are presented below.

Strategies and policies of policy makers

Within the Regional Operational Programmes (1994-1999) for Korinthia and Fthiotis more or less the same strategies are intended for achieving the main objectives of improving the standards of living of the population and the quality of life and reducing disparities. These strategies are: improvement of infrastructure, strengthening of SMEs, further development of tourism (by exploiting its natural, historical and cultural resources), a restructuring and modernization of agriculture, attraction of investments (by providing economic and social infrastructure), protection of the environment and a more effective management of natural resources (forest, water and sea resources).

The mixture of policies available to create a positive environment for employment growth is similar in both regions. However, regional incentives for attracting investment creation in Fthiotis are stronger (15-50% subsidy on investment) compared with Korinthia (only 15% subsidy), as they are derived from the Regional Development Law (1892/91). These have contributed to more investment activities in Fthiotis under this Investment Law. However, total private investment during the period of analysis was higher in Korinthia than in Fthiotis, despite stronger policy incentives in the latter (Katochianou *et al.*, 1998). The most important policy in both regions is related to investment incentives for the creation of new enterprises and the expansion of existing ones, both in industry and tourist activities.

Labour market policies constitute a second important field with regard to direct employment creation. These refer to subsidization of enterprises, in order to create new permanent jobs and to subsidization of the unemployed in order to undertake entrepreneurial initiatives and become self-employed. Although Fthiotis showed lower rates of employment creation by these labour market policies, its vocational training through the apprenticeship programme was more effective than in Korinthia. The improvement of infrastructure – like the highway improvement with support of the Cohesion Fund – has temporary employment effects in construction. At a more permanent base, such improvements reduce travel time and business sites near highways are attractive places for firm settlement. Most delivered policies are at sectoral level, and only recently a LEADER II programme was initiated in Fthiotis' and in Korinthia's mountainous areas.

The modernization of agriculture, which constitutes the main objective of agricultural structural policies, implies a release of agricultural labour in the long run. However, in both regions it also aimed at retaining young farmers in the sector. It is useful to identify policies with a negative effect on employment as well. The CAP reform pressures on farm incomes, due to restrictive supply and price policies, have 'pushed' the farming population out of agriculture at faster rates in Fthiotis than in Korinthia. This was mainly due to the fact that agricultural production in Fthiotis (mainly cereals, cotton, tobacco and olives) is more CAP dependent than the production in Korinthia (mainly grapes, quality wines, citrus, fruits etc.). Besides, the lower level of farm incomes in Fthiotis accentuated its larger agricultural exodus.

The capacity of policy makers (mechanisms, planning and implementing) to formulate strategies and effective policies is weak and this can mainly be attributed to the historical lack of de-centralization that existed in Greece. However, this changes progressively and more power is transferred to regional authorities. These changes are more effectively realized in Korinthia than in Fthiotis.

Strategies of firms towards maintaining or augmenting employment

Enterprises are directly affecting employment, since they are the main source of labour demand. In both regions, almost the total labour force is hired from the local labour markets. In Korinthia, the mix of large and small mainly family-based enterprises forms a very flexible business network, which contributes significantly to employment increases. Strategies of entrepreneurs seem to differ in the study regions. Due to the entrepreneurial tradition in Korinthia, entrepreneurs better perceive opportunities and respond to changes by formulating strategies. This is reflected in the creation of new – mainly small – enterprises, as a response to the crisis in traditional industries. Moreover, due to the industrial tradition, firms in Korinthia are more effectively involved in external networks (exports). On the contrary, the lack of an entrepreneurial tradition in Fthiotis constrains entrepreneurial initiatives and hampers employment growth. Most industrial and service activities refer to small- and medium-sized enterprises, which are in a transformation process towards more competitive conditions. The few large firms in Fthiotis are of exogenous origin and are exporting to international markets. Due to a lack of a tourism strategy in Fthiotis, its employment in the tourist sector is declining, while it is growing in Korinthia.

Farmers' strategies

In both study regions, farmers' strategies observed were:

- a professionalization strategy, which engages more resources in farming and maintains or augments employment in agriculture;
- a disengagement strategy, which removes resources and employment out of farming;
- and a stable reproduction strategy, which maintains the same resources in farming (Arkleton Trust, 1992).

In Korinthia, over half of the farmers (54%) are dynamically involved in farming and tend to pursue the professionalization strategy, while only 12% are disengaging resources or employment out of farming. On the contrary, in Fthiotis less than one-third of the farmers have a professionalization strategy, while another third are disengaging resources out of it. These strategies are consistent with the higher rates of exodus from farming in Fthiotis, compared with Korinthia. These strategies are mainly attributed to differences in agricultural structures prevailing in the case study regions and affect the exodus from farming and overall employment changes.

Another farm adjustment strategy observed widely in both study regions was pluriactivity (Arkleton Trust, 1992). However, pluriactivity has been observed more often in Fthiotis than in Korinthia, due to the less favourable farm structures in Fthiotis, as these push farmers into supplementary employment and incomes (Efstratoglou, 1990). Pluriactivity in both regions is related to off-farm activities such as services, tourism, agricultural processing units, etc. On-farm pluriactivity like agrotourism, or on-farm processing or retailing is not often observed. An explanation for this is that farm households prefer stable income employment to undertaking the risk of on-farm activities. Landscape conservation is not practised, while only recently LEADER II programmes have been implemented in both regions around agrotourism potentials.

Strategies of labourers with regard to employment

Labourers' attitudes are very flexible in both regions due to the prevailing small business sizes and flexible labour markets. The main objectives of representatives of labourers in both regions are the strengthening of labour unions and the intervention through labour agreements in order to reduce unemployment. The expansion of employment contracts to all employed and the emphasis on training are other important strategies. Although there is a demand for initial training in Korinthia, its proximity to Athens has not permitted the development of well-established training structures. Training in Fthiotis seems better developed and covers local needs in a better way. However, in Fthiotis the decline in industries and tourism has made the problem of unemployment more acute, which continues to increase. Although training is considered to be an important strategy for helping labourers to adjust themselves to a rapidly changing environment, it is insufficient as a strategy to create new jobs. An overall development strategy to pursue employment creation through increases in labour demand, is thought to be necessary by labour unions.

Internal/external networks and their impact on employment

Networks, according to their orientation, are distinguished into internal and external ones. In both regions the main internal networks are those of social actors such as Agricultural Co-operatives and their Union, Chambers of Industry and Commerce, Hotels' Association, Labour Union, Young Farmers' Association etc. Regional and local authorities (first and second-degree local government) as well as national authorities constitute another network. In both regions the functioning of internal networks is hampered by little interaction among actors and lack of cooperation among sectors. Due to these weaknesses, the positive impacts of these networks on employment are marginal. There is a need for a more effective cooperation and for building mechanisms and channels that will expand networks and make them more effective. Decentralization and the strengthening of the local government's role may contribute to capacity building in the region, to local partnerships creation and to bottom-up rural development.

External networks refer mainly to administration networks with national and regional authorities and exporting-importing networks of agricultural and industrial products, tour operators and hotel networks. Although external networks in Korinthia are stronger than those in Fthiotis, as actors in Korinthia are traditionally more open and responsive to changes due to the region's proximity to Athens, in general, the interaction of internal and external actors is not very efficient, so there is room for further development and strengthening of this interaction.

Endogenous and exogenous forces behind employment dynamics

It seems that endogenous and exogenous forces have influenced the development of both regions; however, endogenous development forces in Korinthia are stronger compared with those in Fthiotis. More analytically, in trying to identify the role of endogenous/exogenous forces behind employment creation, it can be said that employment growth is a mix of both endogenous and exogenous forces in the regions. In Korinthia exogenous forces played a considerable role during the 1960s and 1970s, but during 1980s and 1990s it seems that endogenous forces had the leading role. This is related to the demographic dynamism and the increasing importance of services, which are mainly endogenous. In Fthiotis, exogenous forces had the leading role during the restructuring period (investment incentives and regional policies in the late 1970s) and seem to continue to play a key role in industrial activities, most of which continue to be exogenously generated. But at the same time, considering the importance of services, it can be seen that endogenous forces are gaining an important role as well. Although endogenous and exogenous forces' impact can not be isolated, it can be argued that both types of force contribute to employment creation, in both regions, but endogenous forces seem to be stronger in the case of Korinthia.

Key factors of success and decline in employment dynamics

Key factors of success in Korinthia

(1) Proximity to Athens

The region's proximity to Athens significantly influenced its development pattern, as during the 1960s a process of economic and social restructuring started, due to the 'spill over' effect of industrial overconcentration in Attica. As saturation points were reached in Attica, industrial activities began to move towards the northern part of Korinthia and were later on diffused to other parts of the region. Because of this, Korinthia has a higher presence of industrial activities compared with most other rural regions.

(2) Smaller exodus from farming

Especially in the level areas, Korinthia's agriculture is intensive with relatively high farm incomes: both value added per hectare and per employee is twice the national average. This has contributed to a slower exodus from farming relative to other rural regions in Greece.

(3) Industrial base of the region

The early industrial development provided a wide range of job opportunities, which contributed to significant population increases, mainly through immigration into the region. Simultaneously, population in rural areas was maintained because of the diffusion of industries to the rural parts of the region.

(4) Demographic dynamism

The increase in population was a key factor behind services' expansion, which were mainly provided by small sized enterprises.

(5) Higher capacity of local actors

Local actors' capacity to perceive opportunities and adjust to changes seems to be higher than in Fthiotis. This is reflected in entrepreneurial dynamism, in developing strategies in tourism and farming and in the presence of more external networks. This open attitude of local actors and their capacity to respond to changes can be attributed to the regions' proximity to Athens.

Key factors of decline in Fthiotis

(1) Tourism decline

The region did not succeed in developing a tourism strategy for exploiting its rich tourism resources: it failed to upgrade tourist services, to promote its local identity and advertise its tourism resources. Hence, the region lost its comparative advantage in tourism.

(2) Faster exodus from farming

Agriculture in Fthiotis is highly dependent on CAP crops like tobacco, cotton, cereals and olive oil. In addition, farm income and productivity per hectare in Fthiotis is relatively low compared with Korinthia, due to more extensive farming systems. The declining CAP support and the negative prospects of farm incomes 'pushed' farmers out of agriculture. This exodus occurred at a higher rate than in Korinthia.

(3) Industrial employment decline

The region has no tradition of industrial production. Industrial development started only in the 1970s – due to the Investment Incentives Scheme – in branches around the regions' local resources like textiles, food processing industries, mining and non-metallic mineral industries. However, the opening to more competitive conditions as a result of Greece's entrance to the EU (1981) affected the industrial development in a negative way and industrial employment decreased. Korinthia, with its longer industrial tradition, showed a higher ability to face the increased competition.

(4) Lower capacity of local actors

Local actors seem to be less flexible, show a more 'dependent' attitude and have a low ability to perceive and adapt to changing conditions. The capacity of policy makers to formulate strategies and effective policies is rather weak. The absence of an industrial tradition constrains entrepreneurial dynamism. The low capacity of actors also hampers the functioning of internal networks, which mostly work on an individual base. The external networks are also characterized by a less efficient interaction of local and external actors.

Concluding remarks and lessons for employment creation

The main goal of this chapter was to present and identify factors behind the different economic performance of two quite similar rural regions in Greece: Korinthia and Fthiotis. From the analysis above it can be concluded that the different rates of population growth, the more favourable farm structures, the more favourable industrial mix, the different dynamism of the service sector and above all differences in capacity of the local actors to exploit opportunities, to develop strategies and to establish more effective networks seem to be the most important factors behind employment dynamics.

Key lessons for employment creation refer to the need to strengthen the capacity of local actors, to develop internal networks and to establish effective links with external actors.

References

Arkleton Trust (1992) *Farm household adjustments in Western Europe 1987–1991.* Final Report on the research programme on farm structure and pluriactivity, Commission of the European Communities, Brussels.

Athanasiou, L., Kavvadias, P., Katochianou, D. and Tonichidou, P. (1995) *Intra-regional analysis and basic data on regions and prefectures (in Greek).* Centre of Planning and Economic Research, KEPE, Athens.

Athens University of Economics and Business (1996) *Labour Market Studies. Greece.* Report for the European Commission, The Research Centre, Athens.

Beatty, C. and Fothergill, S. (1996) Labour market adjustments in areas of chronic industrial decline: The case of the UK coal fields. *Regional Studies* 30, 627-640.

Efstratoglou, S. (1990) Pluriactivity in different socio-economic contexts; A test of the pull-push hypothesis in Greek farming. *Journal of Rural Studies* 6-4, 407-413.

Efstratoglou, S., Mavridou, S. and Psaltopoulos, D. (1995) *Context factors influencing farm women labour conditions. Study area of Fthiotis, Greece.* EU Research Programme DEMETRA, Agricultural University of Athens.

Efstratoglou, S., Efstratoglou, A. and Kalemidou, G. (1998a) *Agricultural and employment in rural areas (RUREMPLO); Case study: The prefecture of Fthiotis, Greece.* Agricultural University of Athens.

Efstratoglou, S., Efstratoglou, A., Kalemidou, G. and Kourousi, E. (1998b) *Agricultural and employment in rural areas (RUREMPLO); Case study: The prefecture of Korinthia, Greece.* Agricultural University of Athens.

European Commission (1996) *Meeting the employment challenge – issues and policies.* European Union Employment Report, Brussels.

European Commission (1997) *Greece; Institutions, procedures and measures.* Mutual Information System on Employment Policies (MISEP), Brussels.

Frost, M. and Spence, N. (1991) British employment in the eighties: The spatial, structural and compositional change of the workforce. *Progress in Planning* 35, 75-168.

Institute of Labour (1999) *Employment and the Greek economy (in Greek).* Annual Report, Athens.

Katochianou, D., Tonikidou, P. and Kavvadias, P. (1997) *Basic data of regional socio-economic development in Greece.* Centre of Planning and Economic Research CEPE, Athens.

National Statistical Service of Greece (NSSG) (1981, 1991) *Population Census 1981, 1991.* Athens.

NSSG (1995) *Labour Force Survey 1995.* Athens.

NSSG (1996) *Statistical Yearbook 1996.* Athens.

OECD (1993) *Creating rural indicators for shaping territorial policy.* Paris.

Overbeek, G., Efstratoglou, S., Saraceno, E. and Haugen, M. (1998) *Labour situation and strategies of farm women in diversified rural areas of Europe.* European Commission, CAP studies, Office for Publications of the European Communities, Luxembourg.

Comparison of the Spanish Case Studies: Albacete and Zamora

7

Jordi Rosell, Lourdes Viladomiu and Gemma Francés

Introduction

According to the OECD index of rurality (OECD, 1994), 20 of the 50 NUTS3 regions in Spain (which correspond to provinces in administrative terms) are classified as 'most rural'. These regions have a very low population density and are generally found in the interior and in the mountainous regions of the Pyrenees and the Cantabrian chain. For the period 1980-1995, three of these 'most rural' regions show an evolution of non-agricultural employment which puts them in the sub-group of 'leading' rural regions; of these three, only Albacete showed an increase in population. Due to the bad performance of non-agricultural employment in this period, Zamora is one of the seven 'most rural' regions or provinces characterized as 'lagging' (Fig. 7.1). In the period 1980-1995 non-agricultural employment in Albacete grew at around 1% above the national rate, whilst in Zamora it grew by 0.2%, around 1% below the national rate (Table 7.1). The province of Albacete is located in the southeast of Spain and borders the so-called Mediterranean Arc, a region of strong economic growth stretching from Girona in the northeast to Malaga in the south. The province of Zamora borders with Portugal and the region of Galicia in the northwest of Spain.

Albacete and Zamora remained on the margins of the process of industrialization, which unfolded in Spain in the second half of the twentieth century. In all, Albacete could count on a few minor nuclei of industrial and

Table 7.1 Evolution of non-agricultural employment and total population growth, 1980-1995 (% per annum)

	Albacete	Zamora	Rural regions	Spain
Non-agricultural employment	2.1	0.2	1.1	1.2
Population	0.6	-0.2	0.2	0.5

Source: INE, Encuesta de Población Activa y Censos y padrones de habitantes.

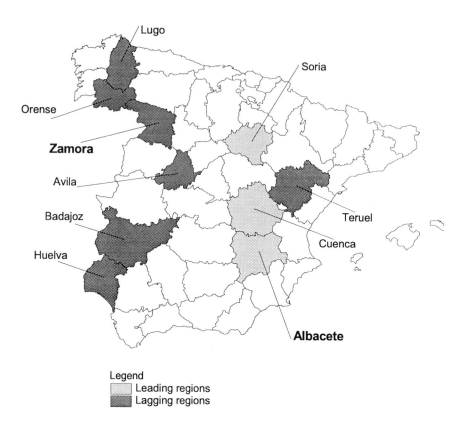

Legend
Leading regions
Lagging regions

Figure 7.1   Leading and lagging most rural regions in Spain

artisan activity: knives in the capital, wine production in Villarobledo and shoes in Almansa. In Zamora, meanwhile, the development of such activity has been far more limited (apart from the production of blankets in the capital and wine production in Toro).

Under the Franco regime, industrial development in Zamora was hampered by the closure of the border with Portugal and the diversion of economic aid away from the region. Neither Albacete nor Zamora benefited from regional development policy, nor was either chosen as the base for any of the large companies investing in Spain. Consequently, both provinces have experienced waves of out-migration throughout this century, but particularly during the 1960s and 1970s as a result of the economic expansion of Spain's major cities: Madrid, Bilbao and Barcelona.

During the period under consideration in this study (1980-1995), Spain experienced a significant change both politically and institutionally. After the death of Dictator General Franco (November 1975) and the celebration of the first free elections (June 1977), the new Spanish constitution was approved at the end of 1978. This initiated a process of increasing regionalization, with the creation of

17 Autonomous Communities to which competences were devolved from the central government. At the same time, there was a notable increase in the financial resources managed by the public administrations and in public sector employment; in addition a welfare system was implemented, albeit somewhat late compared with the levels of welfare achieved by then in the rest of Western Europe. Furthermore, in 1986 Spain acceded to the European Community and set about transposing the range of Community policies. In terms of economic growth and the evolution of employment, after a period of declining GDP and falling employment between 1980 and 1984, the mid-1980s saw the start of a vigorous process of growth which lasted until 1992. Then a period of recession followed, which lasted until 1997.

In administrative terms, Albacete and Zamora are both provinces. Albacete is in the Autonomous Community of Castilla-La Mancha (a NUTS2 region) and Albacete is its largest city, despite the fact that Toledo is the capital. Zamora is in the Autonomous Community of Castilla y Leon (also a NUTS2 region), the largest administrative area in Europe. Valladolid is the regional capital and also its largest city. Both Albacete and Zamora have a per capita income below Spanish and EU averages (Table 7.2). In addition, both are classified as 'Objective 1' areas under EU structural policy. Furthermore, the Spanish government has designated both provinces as Areas for Economic Promotion, making them eligible for funding from the Interterritorial Compensation Fund; this was created in 1984 with the objective of financing investment in Spain's least developed regions. There are also EU Leader II programmes in operation in both provinces. As such, there is ostensibly no great difference in terms of the financial resources received by the two provinces to promote their economic development. In both cases, the bulk of funding received from the EU (via ERDF, ESF and Cohesion Fund) is administered by the regional authorities.

Albacete has just over 360,000 inhabitants in an area of almost 15,000 square kilometres, giving a population density of 24 inhabitants/km^2 (Table 7.2). A little over one-third of the provincial population lives in the city of Albacete, and the province has three other important urban centres, each with a population over 20,000 (Hellin, Almansa and Villarobledo) (Table 7.4); together, these four show a notable population increase over the period of study. Conversely, the province's mountainous southwestern areas (i.e. Sierra del Segura and Sierra de Alcaraz counties) have very low population densities (barely 8 inhabitants/km^2) and are subject to ongoing decline in population.

Zamora has just over 200,000 inhabitants across an area of 10,500 km^2, giving a population density of barely 19 inhabitants/km^2. The city of Zamora is the province's largest, with 66,000 inhabitants; there are no other urban centres with more than 20,000 inhabitants. Only in Benavente (16,000 inhabitants) a certain degree of services is available. The province's western boroughs, those which are mountainous and border Portugal (i.e. Sanabria, Aliste and Sayago), have very low population densities (fewer than 10 inhabitants/km^2) and moreover, are subject to ongoing depopulation.

Both Albacete and Zamora experienced considerable out-migration between the 1950s and the mid-1970s, resulting in a significant population decline. In the

Table 7.2 Economic environment of Albacete, Zamora and Spain

	Albacete	Zamora	Spain
Population, 1995	361,300	204,800	40,322,200
Size (km²)	14,900	10,600	506,000
Density (inh./km²)	24.2	19.4	79.7
Population growth			
1980-1995 (%)	8.9	-4.7	7.7
Employment growth			
1980-1995 (%)	3.6	-33.1	5.9
Sectoral employment			
structure, 1995 (%)			
Agriculture	11	24	9
Industry	30	23	30
Services	58	53	61
Population over 65 years (%)			
1980	12	18	9
1995	14	22	14
Rate of unemployment (%)			
1980	12	9	12
1996	26	18	22
Income per capita (index, EU=100)			
1980	53	56	70
1996	66	65	76

Source: INE.

Table 7.3 Population in main cities

Albacete		Zamora		Spain	
Albacete	130,000	Zamora	66,000	Madrid	3,029,700
Hellín	24,200	Benavente	15,900	Barcelona	1,614,600
Almansa	22,600	Toro	9,700	Valencia	736,300
Villarrobledo	20,700			Sevilla	719,600

Source: INE.

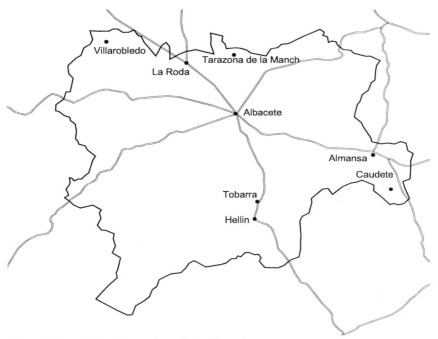

Figure 7.2   Main cities and roads in Albacete

Figure 7.3   Main cities and roads in Zamora

mid-1970s, coinciding with the national economic crisis, demographic trends stabilized in Albacete. The outflow of population all but stopped. Thereafter, Albacete's levels of in-migration were favourable. The same has not been the case for Zamora: although the economic crisis of the 1970s considerably reduced out-migration, the province has continued to lose population. At the start of the 1980s, the population of Albacete and, in particular, of Zamora showed higher levels of ageing than the national average (Table 7.2). Nevertheless, the under-25 population of Albacete still constituted 44% of the provincial total (compared with a nation share of 42%), suggesting a relatively youthful population. In the mid-1990s, Albacete's population structure by age was close to the national average, and considerably younger than the average for rural regions, whilst Zamora showed above average levels of ageing.

Local resources

The land area of Albacete is predominantly made up of plains, whilst in the south the landscape becomes mountainous (Sierra del Segura and Sierra Alcaraz counties). The eastern part of Zamora is also very flat, whilst in the west the landscape becomes more mountainous, particularly in the northwest (Sanabria county). Forested areas account for 21% of the total area in Albacete and 14% in Zamora. Utilized agricultural area (UAA) accounts for 64% of Albacete and 66% of Zamora; permanent crops (mostly vineyards) represent 16% of UAA in Albacete, but barely 2% in Zamora. Pastures cover 30% of the UAA in Zamora and less than 18% in Albacete; livestock production is more important in the agricultural production of Zamora (44%) than in Albacete (15%). In both provinces, most of the UAA is designated as Less Favoured Area (LFA): 72% in Albacete and 69% in Zamora.

Part-time agriculture is increasingly common in Albacete, but less in Zamora, where 82% of farm owners continue to work full-time on their farms. In Zamora, 28% of farms are less than 5 hectares in size, compared with 78% in Albacete. Despite the large number of small farms, about 3.5% of farms in Albacete are over 200 ha, accounting for 50% of the UAA. In Zamora, agriculture is mainly low-input, rain-fed cereal cultivation offering yields close to the national average.

Both areas have low and irregular rainfall, and in addition extreme temperature ranges seasonally as well as between day and night. Given these climatic conditions, the traditional crops in Albacete and Zamora are rain-fed winter cereals and permanent crops. In recent decades, the area of irrigated land – using both surface and ground water – has increased. The increase in irrigation has been larger in Albacete than in Zamora. Irrigation allows the introduction and development of more profitable livestock and crop varieties, such as maize, sugar beet, alfalfa, vegetables, malt maize, mushrooms, garlic and melon. Irrigation became the basis for new industries and services as well. Moreover, the climate of Albacete is less extreme than that of Zamora, particularly in winter, and this has favoured the agricultural diversification process. Nevertheless, concerns exist at the over-extraction of water from the aquifers.

The food industry is the most important industrial sector in Zamora and the second most important in Albacete after textiles and footwear. In both provinces, agro-industrial businesses tend to be associated with the traditional products of wine, cereals, milk and meat. In the case of Albacete, businesses have diversified and now also sell their products across the whole of Spain, whilst Zamoran products continue to be sold principally via local and regional markets. Zamora is home to one of Spain's most important hydrological networks, so that hydroelectric production is the most important local resource. There are 13 reservoirs devoted to the generation of hydroelectricity, with a combined capacity of around 3.5 million megawatt hours (MWh) and exports totalling 2 million MWh. However, this provides employment for only 40 people in Zamora: the generating company, Iberdrola, has its headquarters in Bilbao in the Basque Country, and it is here that it pays its taxes and where most of its staff is employed.

Neither Albacete nor Zamora has mineral resources of any importance. In Albacete, certain enterprises require the availability of natural resources: mineral water, dolomites for paint production in La Roda, as well as a particular clay for use in the construction sector. In Zamora, the other local resources apart from hydroelectricity are granite, slate and some clay and limestone; these resources have only recently started to be exploited.

Transport networks in both provinces have been improved in recent years. The most significant development has occurred in Albacete: while in 1980 there was not a single kilometre of motorway in the province, 15 years later there were over 150 kilometres of motorway, linking the province to the country's main economic centres and corridors. The new motorway between Alicante and Madrid and Almansa and Valencia constitutes the principal transport link between Madrid and the Levant region of eastern Spain. This motorway has had a 'connecting corridor effect'. Nowadays, the counties and municipalities located on the Alicante-Madrid axis (Comarca de Almansa, Los Llanos and Mancha Albacetena), as well as, to a lesser extent, on the Murcia-Madrid axis (Campos de Hellin), enjoy easy access to Spain's most dynamic economic axis, the Mediterranean Arc, and the urban-industrial centre of Madrid. This infrastructural improvement has reduced travel time to the important markets of Madrid, Valencia and of the Levant in general (i.e. the ports of Cartagena and Alicante and the airports of Alicante and Valencia). As such, it has undoubtedly enabled the economic dynamism of the Mediterranean Arc to spread to certain counties in Albacete. In Zamora, a motorway crossing the northeastern part of the province (Benavente) links it with Madrid. This motorway is supposed to continue to Galicia, but progress has been slow and it has yet to be completed. Considerable difficulties persist, not least of which is the connection with Portugal.

In parts of Albacete such as Sierra del Segura, Sierra de Alcaraz, Campos de Montiel and La Manchuela, landscapes, lakes, springs and mountains are important in the development of the tourism sector. However, this kind of tourist resource in Albacete should be considered as of a secondary order and of less interest compared with that of its neighbouring coastal provinces. Zamora has a number of natural tourist attractions, such as the Lake Sanabria Natural Park, the Villafafila Lagoons Natural Reserve, the Sierra de Culebra and Los Arribes del

Duero. In addition to these natural areas, Zamora has an architectural heritage, including Roman architecture and mediaeval castles, which are seen as important components in the development of a new type of tourism. However, tourism infrastructure is underdeveloped, and the number of overnight stays in hotels, hostels and the like has not increased significantly over the period of study (220,000 in 1980 and 240,000 in 1996) (INE, Statistical Yearbooks).

Economic activities

According to official data[1], economic growth for the period 1980 to 1993 in both provinces was below the national average. GDP in Albacete grew at an annual rate of 2.1%, against the national rate of 2.5%. In Zamora, GDP grew at an average annual rate of 1.7% over the same period, and as such was below the average of the group of most rural regions. The evolution of employment in each province was quite distinct (Table 7.4). For the period 1980-1995, there was a marked decline in agricultural employment in both provinces. However, of greater significance has been the evolution of non-agricultural employment. In Albacete, industrial employment grew by 0.6% per year, against a decline of 1.4% in Zamora, which was well below the national rate as well as that for the group of most rural regions.

In which industrial activities did employment in Albacete increase? What happened over the same period in Zamora? According to the disaggregated BBV (1997) data, between 1983 and 1993 about 5,700 new jobs were created in the secondary sector of Albacete and 2,900 in that of Zamora. In Albacete, two out of every three new jobs in the secondary sector were created in the manufacturing

Table 7.4 Employment growth by sector, 1980-1995 (in % per year)

	Total	Agriculture	Non-agrarian	Industry	Services
Albacete	0.2	-6.6	2.1	0.6	3.0
Zamora	-2.2	-5.8	0.2	-1.4	0.8
Most rural regions	-0.8	-5.3	1.1	-0.3	2.0
Spain	0.4	-4.7	1.2	-1.0	2.4

Source: INE, Encuesta de Población Activa.

---

[1] The statistical data used here were derived from official sources (*Instituto Nacional de Estadística*/National Statistics Institute (INE)) and from the Research Department of the Bank of Bilbao Vizcaya (*Banco Bilbao Vizcaya*). Data on the evolution of Gross Domestic Product (GDP) are from the INE Regional Accounts Department (INE, 1997) and cover the period 1980 to 1993. Employment data for the three main sectors (agriculture, industry and services) are taken from the INE Labour Force Survey and cover the period 1980 to 1995. Finally, the more disaggregated data concerning the evolution of employment in both regions are taken from the biannual publication *Renta Nacional de España y su distribución provincial*/National Income in Spain and its Provincial Distribution, produced by the Bank of Bilbao (BBV, 1997). Unfortunately, these data are only available for the years 1983 to 1993 and show sometimes opposite trends compared with INE data.

industries, which thus grew from 16,300 to 20,300 jobs and showed a total increase of 24% over the period. The majority of these almost 4,000 new manufacturing jobs were created in three traditional activities – metal products, textiles and footwear and food and drink. By 1993, manufacturing activities accounted for almost a quarter of non-agricultural employment. In Zamora, barely 600 new jobs were created in the manufacturing industries, equal to an increase of 12%. In 1993, manufacturing accounted for scarcely 12% of non-agricultural employment.

The remaining jobs in the secondary sector were created in construction: a little over 2,200 new jobs in Albacete and a further 2,750 in Zamora. In this latter province, the construction boom was especially noteworthy: employment in this sector grew by 64% between 1983 and 1993 (compared with 36% in Albacete) and came to represent 15% of non-agricultural employment (compared with 10% in Albacete). Employment in mining declined in both provinces, even though it was a sector of little economic importance. Finally, in water, gas and electricity supply, barely 40 jobs were created in Albacete, whilst 250 were lost in Zamora as a result of the growing automation of the province's hydroelectric installations.

Jobs were created in the service sector across the whole of Spain (Table 7.4). In Albacete, service sector growth was particularly marked and occurred at a higher rate than the national and rural averages. In Zamora, meanwhile, growth in the tertiary sector was poor. The majority of new jobs were created in public services: 7,600 new jobs in Albacete and 4,600 in Zamora. Public sector growth in Albacete between 1983 and 1993 was somewhat above that in Zamora (70% compared with 65%). However, the strong performance in tertiary employment in Albacete compared with the weak performance of Zamora can be explained by what occurred in the private service sectors of the respective provinces. In Albacete, 4,600 jobs were created in retail trade, hotels and restaurants, this representing an increase of 33% between 1983 and 1993, whilst in Zamora, the 2,500 new jobs created in these sectors constituted an increase of 25%. The difference is greater in those tertiary activities referred to as services to business (i.e. transport and communications, credit and insurance etc.), which grew by 20% in Albacete and by 10% in Zamora. Table 7.5 shows those sub-sectors which enjoyed a net increase of over 500 employees between 1983 and 1993. It can be seen that there was greater industrial dynamism in Albacete than in Zamora, and that the service sector also performed better there.

The evolution of the apparent productivity of labour and its comparison with the evolution of employment helps to elucidate the economic and productive trajectories of both provinces. Albacete demonstrates an above average increase in employment but not in productivity in the period 1980-1993, whilst Zamora is below average in both (Fig. 7.4). The Albacete pattern is common in 'most rural' regions of Spain, and might be characterized as 'restructuring via employment'. That is to say, a growth in economic activities which are labour intensive. This is especially clear in the case of industry, where productivity even declined. In fact, the growth in manufacturing employment in Albacete occurred in traditional industries with low productivity. Conversely, the experience of Zamora has been that of a 'vicious circle', where performance in both employment and productivity has been below the national average. The strong performance of industrial

productivity in Zamora is worth noting, however, this can be accounted for by the
high levels of productivity achieved in hydroelectric power generation.

Table 7.5 Branches with high employment increase, 1983-1993

	Albacete	Zamora
Manufacturing	+++	+
Metal products and machinery	++	+
Textile and footwear	++	
Food and drink	+	
Construction	++	++
Services	++++	++++
Public services	++++	+++
Wholesale and retail trade	++	++
Hotels and restaurants	++	+
Transport and communication	++	
Household services	++	

Note: 501-1,000 new jobs = +; 1,001-3,000 new jobs = ++; 3,001-5,000 new jobs =
+++; 5,001-10,000 new jobs = ++++.
Source: BBV, 1997.

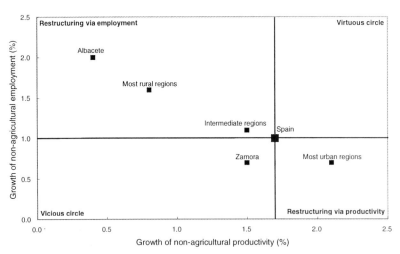

Figure 7.4   Relationship between non-agricultural productivity and employment,
1980-1993
Source: INE, Contabilidad Regional de España 1983, Madrid 1985.

Actors

In the analysis of actors' strategies, let us now look at the conduct of farmers, entrepreneurs and workers. Within the latter group, special reference will be made to the experience of women. Thereafter the strategies of policy makers are analysed, and finally some observations are made on actor networks and their dynamics.

Farmers

In Albacete, land tenure is concentrated in very large properties which exist alongside small holdings of only a few hectares. These large farms, which are operated by a paid workforce, have over time been converting their land to irrigated systems, mostly using groundwater, and switching to products with a higher value added such as maize, alfalfa and sugar beet. This change has not brought about an increase in employment, given the mechanization of irrigation and of crops. The long-term sustainability of water supplies from the underlying aquifers is seriously in question. The strategy of small farmers has been to pursue pluriactivity as well as intensive production on small irrigated plots, cultivating products with a high value added and requiring a high labour input: garlic, mushrooms, melons, saffron, fruit trees and the like.

In Zamora, agricultural land is more equal distributed amongst small and medium-sized holdings. Farmers who have converted to irrigated agriculture have mainly continued to produce cereals. At the same time, farmers and shepherds have traditionally constituted two distinct groups. Shepherds were not landowners, but in the past few years they have been acquiring plots of farmland. The most visible strategy of landowning farmers is on the one hand abandonment, and on the other hand, expansion of farmed land as a means of securing farm income exclusively from agriculture. Abandonment occurs when there is no successor, as many farmers send their sons and daughters off to study outside the province, after which they do not return. Expansion of the farmed area is achieved by renting out additional plots, since farmland comes only rarely on the open market.

Compensatory payments under the CAP have served to improve the financial situation of farmers in both Albacete and Zamora.

Entrepreneurs

Using an index of the number of workers employed per firm, the two provinces have similar entrepreneurial structures. Approximately 80% of businesses employ fewer than six workers (Table 7.6). In the industrial sector, small businesses are even more prevalent. In 1995, there was only one manufacturing company in Zamora employing more than 100 workers. In Albacete, there were 15 manufacturing companies with more than 100 employees, together accounting for over 2,300 workers (at an average of 155 workers per company). Nevertheless,

Table 7.6 Enterprise structure according to number of workers, 1996 (%)

	Total number of enterprises	Number of workers per enterprise				
		1-5	6-50	51-100	100-500	500+
Albacete	8,649	81.5	17	1	1	0.1
Zamora	5,393	83.5	15	0.5	0.5	0.5

Source: General Treasury of Zamora's Social Security, INEM and INE.

there are some examples of newly established medium-sized enterprises (20-100 workers per company) engaged in commercial distribution, transport and communications, health and financial services.

In fact, small enterprises (those with fewer than 20 employees) have been the principal, indeed almost the only, source of non-agricultural jobs in the private sector, while the few medium-sized and large companies have either disappeared or have reduced their workforce. In Albacete, and to a lesser extent in Zamora, the larger industrial companies (those involved in the production of footwear, textiles, clothing and the like) were the worst affected by the economic crisis of the 1970s and early 1980s, and many disappeared. In Albacete's most dynamic boroughs, new businesses have been set up by former workers, some of whom previously worked for the large companies which have gone out of business. These new businesses tend to be small, although some have been able to grow, making frequent use of sub-contracting and of home workers. Such businesses often come together to form 'industrial districts', taking advantage of positive externalities (such as a trained workforce, ease of supply and the availability of infrastructure) as well as economies of agglomeration. In Zamora, with a weaker manufacturing tradition, this type of business formation has been far less prevalent. New businesses, or rather micro-businesses, have tended to be set up in the construction and service sectors. In addition, a number of small workshops have been established to produce metal goods.

This business strategy, which we might refer to as 'self-employment' (securing employment by setting up a business or micro-business), exist side by side with a second strategy which we might call 'survival' (securing continuity of employment in a changing environment). In this connection, there has been a notable trend in Albacete towards a marked specialization in certain market niches (i.e. in terms of products, market segments and the like). This occurred for example in the footwear sector (cowboy boots) and in the cutlery sector (sports knives, decorative cutlery).

In Albacete, reference must also be made to a strategy of 'consolidation', corresponding to certain medium-sized businesses in the food, metal products, transport and chemical sub-sectors. This strategy has been pursued by companies which have grown considerably and which have thereby been able to extend their markets beyond the region and moreover to implement a series of technological improvements.

In both Albacete and Zamora, local individuals set up the majority of new business initiatives. However, Albacete, unlike Zamora, has also been able to attract a good number of businesses from outside the province, thanks to its geographical location and to the fact that it is able to offer subsidies to investment (i.e. regional incentives) as well as lower wage rates.

Labourers

The incorporation of women into the labour market has been one of the most noteworthy changes undergone in Spanish society and labour markets in recent decades. For rural regions, where women's participation in the labour market has been and remains lower, this change has constituted a formidable challenge. Indeed, the possibility of creating jobs for women has turned out to be a decisive factor in terms of retaining and even increasing the rural population. Across Spain, the increasing level of women's participation has been notable in the period in question (Table 7.7). This increase has more than made up for the fall in male participation rate, caused, on the one hand, by longer periods spent in education, and, on the other, by the ageing of the population.

In the context of female participation in the labour force, the experiences of Albacete and Zamora have been diametrically opposed. In Albacete female economic activity increased at a much faster rate than the national average. Consequently, at the end of the period in question, Albacete has an overall participation rate (men and women) on a par with the national rate and considerably above the average for rural areas (43%). The strong increase in female employment might partly explain the low productivity growth in Albacete, as women are relatively more employed in unskilled professions. The experience of Zamora has been exceptional: there has been a fall in the female participation rate, which, at the end of the study period (despite a partial recovery), stands at ten points below the national average.

Table 7.7 Participation rate by gender in Albacete, Zamora and Spain, 1980-1995 (%)

	Total			Men			Women		
	Albacete	Zamora	Spain	Albacete	Zamora	Spain	Albacete	Zamora	Spain
1980	44	48	48	69	68	72	20	28	27
1985	44	43	47	66	63	69	21	23	28
1990	49	41	49	67	60	66	32	23	33
1995	49	39	49	61	52	63	36	26	36

Source: INE.

Table 7.8 Employment levels by gender in Albacete, Zamora and Spain, 1980-1995

	Men			Women		
	Albacete	Zamora	Spain	Albacete	Zamora	Spain
1980	75,970	54,810	8,170,770	21,200	21,680	3,263,650
1995	68,080	37,450	7,920,230	32,500	17,480	4,222,430

Source: INE, Encuesta de Población Activa (several years).

This highly distinctive evolution of the participation rate reflects what has happened in terms of employment. In Albacete, there has been an increase in the level of female employment: the number of women employed has increased by 11,300 between 1980 and 1995 (Table 7.8), equivalent to a 3% annual growth rate. The majority of jobs created in the service sector and a large part of those created in manufacturing were for women. Of those activities, which generate considerable employment, only construction offered more jobs for men than for women. The number of women employed over the same period in Zamora has been reduced by over 4,000. This equates to an annual loss rate of -1.4%, in contrast to the national experience (an annual growth rate of 1.7%) as well as to that of rural areas (a modest annual growth rate of 0.1%). The mediocre increase in employment in manufacturing industry and services illustrates the difficulty of creating jobs for women. This incapacity to create female employment is expressed more by out-migration than by remaining unemployed. This explains the fact that there is a higher level of unemployment in Albacete than in Zamora (Table 7.2).

The unemployment rate has increased in both provinces. In Spain as a whole, the average unemployment rate for the year 1995 was 23%, 3% higher than that for Zamora and 2% below that for Albacete. In rural areas, the average rate was 21%. It is important to stress that the level of unemployment in Albacete has risen as a function of the increase in the pool of labour available, whilst in Zamora the level of unemployment has risen at the same time as the pool of labour available has contracted.

Levels of education are higher in Zamora than in Albacete (Table 7.9). In Albacete, 28% of the population has received no education, whilst for Zamora the figure is 18%. Moreover, a slightly higher percentage in Zamora has been through tertiary education and a higher percentage has also received primary education in Zamora. This suggests that in terms of job creation the level of education is not so important, but rather the suitability of such to the requirements of the labour market. Indeed, the majority of the best educated in Zamora eventually emigrates to more dynamic urban centres.

Table 7.9 Levels of education in Albacete and Zamora, 1991

	Albacete		Zamora	
	total no.	share population (%)	total no.	share population (%)
No studies	63,200	28	24,400	18
Primary	64,500	29	50,000	38
Secondary	81,700	36	47,000	35
Tertiary	12,200	5	9,900	7

Source: INE.

Average wage levels in Castilla-La Mancha (which includes Albacete) and Castilla y Leon (which includes Zamora) are lower than those of Madrid, although this gap has closed in recent years. In Castilla-La Mancha, the average fixed wage is around 22% lower than that in Madrid, while in Castilla y Leon it is 13% lower.

The foregoing illustrates the different strategies followed by workers in each of the two study regions. While in Albacete efforts to improve the availability of employment have increased, in Zamora there has been a contraction in the pool of labour. In Albacete, women are being incorporated into the labour market and self-employment and farm family helpers are increasing, as are pluriactivity and a significant informal economy. In Zamora, conversely, there remains a tendency towards out-migration, a continuing aspiration amongst the population to become civil servants, and a willingness to live of income from the land and from the CAP.

Policy makers

Since the establishment of the new Spanish constitution in 1978, a process of decentralization has been unfolding, in which local and regional administrations have a wider range of competencies. Regional incentives are administered by the regional governments (Autonomous Communities). Funds are provided by the EU and by central government. In Zamora, a special Economic Development Agency also administers funds, but it has been settled only in 1997.

Expanding the road network has been a priority in both provinces, the aim being to increase market access and to improve the accessibility of various municipalities. Hence, the bulk of European structural funding (under the ERDF and the Cohesion Fund) has been spent on improving transport infrastructure. The telephone network, sports facilities, schools and medical centres have also been improved. At the same time, policy makers in each province tend to approach job creation in different ways. In Albacete, policy makers have sought to foster job creation and entrepreneurial activity, whilst in Zamora, policy makers have not shown themselves to be overly sensitive to entrepreneurial endeavours. As such, the dynamism, local entrepreneurialism and cooperation networks found in

Albacete present a contrast to the cautious and risk averse disposition of entrepreneurs and policy makers alike in Zamora, dispositions which may have the effect of constraining provincial economic growth. In the Albacete context, it is worth noting the significant efforts made to attract new businesses, offering incentives (lower land prices etc.), creating a favourable environment for the labour market and providing tailored training programmes. One example has been the 'Pacto de Albacete', an agreement similar to the Territorial Employment Pacts promoted by the European Commission. Albacete has been successful in attracting business investment from outside the province, as well as in a few cases from outside Spain. Priority has been given to setting up training programmes for the workforce tailored to the needs of new investment. The number of training courses has risen, and the training programme as a whole has been strengthened by support from the European Social Fund.

In Albacete, the Leader I and II programmes, with two and three Local Action Groups respectively, have facilitated economic diversification by making possible the incorporation of new economic activities such as tourism. In Zamora, despite the province's rurality and the fact that it is seen to be a lagging rural region, no initiatives were established under Leader I.

Analysis of actors' interaction

In Zamora, the most influential civic organizations are the farmers' unions. Industry and commerce entrepreneurs have different organizations among which significant conflicts exist. Leader II initiatives were promoted by the farmers' unions or by a non-profit organization from outside the province. Entrepreneurial activity in Zamora originates locally and is limited to local markets, so that there are no meaningful trade links with other regions or countries. In recent years, there have been attempts on the part of some institutions to broaden relations with Portugal, which had previously been non-existent. Overall, the process is very slow. Political parties are dominated by powerful local leaders, and wrangles between them tend to obstruct the working of the local government. Zamora is an example of socio-economic isolation with only weak development of internal and external networks.

In Albacete, the institutional fabric is better established and there is stronger participation by commercial organizations (employers' organizations, the Chamber of Commerce etc.) and trade unions. In the period 1980-1995, there was fruitful collaboration between the institutions working towards the economic development of the province and the creation of employment. In Albacete, farmers and entrepreneurs are each represented by their respective organizations. More recently, groups have formed that widen the net of local actors. The principal export good in Albacete is footwear. Wine, olive oil and cutlery are also important. Industrial development in the province has occurred in 'industrial districts' which benefit from significant agglomeration economies (for example, footwear production in Almansa, knives in Albacete, paint in La Roda and chair making in La Gineta). At the same time, traditional enterprises persist alongside newly established, more flexible and specialized companies.

Key factors of employment dynamics

Having shared similar experiences during the 1970s, Albacete and Zamora came to follow different trajectories over the period 1980 to 1995. Factors contributing to these diverging trajectories may be summarized as follows:

(1) Regional location

With the creation of the Autonomous Communities, Albacete transformed from a poor, marginal part in the former region of Murcia into the principal urban settlement (i.e. the city of Albacete) in the new region of Castilla-La Mancha. This shift into the most dynamic zone in the region was enhanced by its location adjacent to the Mediterranean Arc. By contrast, Zamora is in a disadvantaged location in the middle of Castilla y Leon. In this weakly industrialized region, economic dynamism and indeed development policy, are concentrated on the Valladolid-Palencia-Burgos axis, to the detriment of the rest of the region.

(2) Degree of isolation

Improvements in the transport and communications infrastructure have been very important for Albacete, which benefits from being at the mid-point of the

Table 7.10   Developmental constraints and incentives in Albacete and Zamora

	Albacete	Zamora
Regional location	Strong	Weak
Demography	Young population	Ageing population Emigration
Communication infrastructure	Connecting corridor: Madrid-Mediterranean Arc	*Cul-de-sac*: Frontier
Urban infrastructure	Good urban structure • Large provincial capital • 3 medium sized towns	Dispersion No big capital nor medium sized towns
General productive structure	Diversification • Manufacturing • Tourism • Services	Predominantly agricultural Dependence on CAP
Industry	Industrial tradition Modernization Niche products	No tradition in industrial development
Strategies of policy makers	Entrepreneurial, developmental outlook Regional subsidies	Local elite
Agriculture	Pluriactivity New agricultural products	Traditional cultivation • Cereals • Sheep

Mediterranean Arc (CPLPSDAMS, 1996) as well as from connections between Madrid and the Levant, the so-called 'connecting corridor' effect. By comparison, Zamora is relatively isolated, since connections with Portugal are poor and so produce a barrier effect (López Trigal, 1994) and since it benefits only indirectly from the Madrid-Galicia axis, which, moreover, has yet to be completed.

(3) Exploitation of local resources

Rural amenities in Albacete refer to beautiful landscapes, springs, lakes and mountains in several parts of the region and in Zamora to historical and cultural heritage like Roman castles and dovecotes and to natural parks. However, these rural amenities do not attract the large number of tourists like in the coastal provinces of Spain, although in recent years inland areas become more popular. For Albacete, the Leader I and II programmes are acting as an important stimulus for its nascent tourist sector. In Zamora tourist attractions lack the infrastructure which would allow their fullest and most profitable exploitation, although there have been certain improvements in more recent years.

(4) Productive diversification

In the 1960s, the economy of both provinces was essentially agrarian, with a degree of artisan industrial activity and some early signs of development in the service sector. Today, Zamora remains basically an agrarian economy, heavily dependent on agro-industry and services associated with the agricultural sector. Furthermore, Zamora is heavily dependent on the CAP. Conversely, Albacete has been able to make progress towards productive diversification, both intersectoral and intrasectoral, exploiting new markets and gaining a strong position in certain niches, and attracting business investment from outside the province.

(5) Utilization of socio-economic policy

Policies to promote socio-economic development and the funds accompanying them are comparable in both provinces, having been implemented according to the European Commission framework as well as to those designed by regional and national government. Differences appear in terms of the degree to which actors in each province have been able to make use of the opportunities on offer. On the one hand, Zamora has been slow to make the most of those instruments best suited to its situation. A salient example is the fact that there are no Leader I initiatives in the province, although the region is clearly one which could benefit and indeed would be eligible to benefit from this programme; a further example is that the Local Development Agency established offices in the province only recently. On the other hand, actors in Albacete, confronted by serious unemployment problems, have competed for development funding to attract capital, create new businesses, improve education, attract public services etc. Zamora, meanwhile, was applying for income transfers and public service improvements which, although they enhanced people's quality of life, failed to generate any employment.

(6) Demography

Both provinces have experienced considerable out-migration as well as ageing of the population. The two processes still persist in Zamora, whilst they have been brought under control in Albacete. Demographic evolution in Albacete is giving rise to new economic demands and expectations. Zamora has entered into a worsening demographic trap, with highly negative effects which are expressed in a contraction in economic activity, a slowing down in the incorporation of women into the labour market, a rapid increase in the number of pensioners etc. Due to depopulation, the scattered villages of Zamora become more and more empty. In order to face this tendency and to create some critical mass, attempts are made to concentrate population and economic activities in the larger villages.

Concluding remarks

Albacete and Zamora are examples of two types of Southern European rural regions. They have to generate new means of development and survival from within, since they are far from Europe's main axes of economic dynamism and tend to be overlooked by politicians and business investors. Clearly, the two provinces differ in many respects: Albacete covers a larger area and has a bigger population, whereas Zamora's location is more peripheral and its population more scattered etc. Nevertheless, the analysis of the trajectories of both study regions suggests different lessons, which we will discuss below.

(1) Agriculture continues to play an important role in maintaining employment

At the end of the 1970s, agriculture in both provinces shared a similar productive orientation: cereals, sheep and vines were the dominant economic activities. In the case of Albacete, a common strategy amongst farmers has been to pursue pluriactivity and diversification of production; this has aided in the generation and maintenance of employment in agro-industry (i.e. the food industry and the production of associated agricultural inputs). In Zamora, on the other hand, farmers have tended to remain engaged full-time in agriculture and to adhere to a restricted range of traditional products; the result being that the number of farmers in the province has been steadily falling. So, the diversification of agriculture has positive effects on direct employment (part-time, flexible) as well as indirect employment (food industry, etc.).

(2) Trends in traditional manufacturing account for the greatest difference

Over the last 15 years, Albacete has expanded and diversified its industrial base. Industrial activity tends to be organized in 'industrial districts', and has exploited market niches and also, for the first time, has attracted capital from outside the province. However, these activities mainly involve traditional products, with low productivity and conditions of strong competition. Conversely, in Zamora, the little industry which existed previously has been steadily disappearing. Industrial expansion in Albacete has had two collateral effects: to create employment for the female workforce and to change the composition of local elites, taking more importance of industrial entrepreneurs.

(3) Public sector has been principal generator of employment in absolute terms

Over the period 1980-1995, all of Spain benefited from a considerable growth in public employment. However, this increase was not evenly distributed across the regions. Albacete experienced a considerable growth in the public sector, particularly in the creation of new public services such as the university, the hospital etc. The creation and location of these jobs in Albacete was a function of changes in the geopolitical map, in which Albacete is the largest city of the Autonomous Community of Castilla-La Mancha, and of the ability of local actors to promote their area.

(4) Availability of a skilled workforce suited to the needs of the labour market

A comparison of education levels in the two provinces shows that levels are higher in Zamora than in Albacete. However, the analysis also suggests that the workforce in Albacete has been better able to match the demands of the labour market. The created jobs in Albacete require various skills as well as appropriate professional training. At the same time, the mismatch of supply and demand of labour in Zamora reinforces the process of youth out-migration, since the large number of Zamora students attending universities outside the province usually do not return.

(5) Strengthening the capacity of local actors and creating strong networks

To a large extent, the success of Albacete can be explained by the fact that, over the past 15 years, its local actors have been able to diagnose the socio-economic situation correctly, to define objectives, to make good use of structural policies and regional incentives and to coordinate their activities. By contrast, Zamora, in our opinion, has remained cut off from mainstream developmental processes, with a traditional society still in place so that farmers have retained their elite position and, at the same time, a policy of seeking assistance rather than development has taken root.

Challenges faced by both regions quite different

In Zamora, the problems typify those of a deep rural area, plagued by negative demographic trends and a progressive ageing of the population. This process is exacerbated by the province's geographic isolation and socio-political marginality. Policies implemented over the period 1980-1995 failed to change the province's outlook. Two indicators demonstrate the gravity of the situation. On the one hand, the level of female employment declined over the period in question, an exceptional experience in the Spanish context. On the other hand, since 1995, the number of pensioners has outstripped the number of contributors to the social security system. Moreover, the analysis has pointed to social and institutional factors which hamper development. The local elite in Zamora is principally composed of farmers, and farm unions play a major role in the institutions of the region. This social group is unwilling to accept the need for change. It defends the maintenance of the status quo and applies for aid policies. It seeks greater direct income payments but not aids for investments or job creation. The 1992 CAP reform reaffirmed these demands whilst at the same time making it easier for beneficiaries to live and work outside the region, thus making life easier for absentee landlords and exacerbating loss of income to the region.

The small number of new Zamoran businessmen interviewed indicated that they had enjoyed very little collaboration in the region on the part of policy makers and institutions. There is little awareness of the need to develop instruments to attract investment and initiatives from outside the province. In sum, without a strong push from outside to facilitate a change in planning and the establishment and consolidation of effective local networks, it would seem that the creation of employment and a change in the demographic dynamic are objectives impossible to fulfil at the present time in Zamora.

In Albacete, the challenge has to do with consolidating jobs, creating more skilled jobs and maintaining a rate of growth in employment which can deal with the rapid rise in the economically active population. Many of the businesses which generated employment in the period 1980-1995 are open to international competition and so highly sensitive to labour costs. These are vulnerable enterprises in which the entrepreneur has to defend and maintain his market niche so as to ensure survival. At the same time, much of the employment created is relatively unskilled, precarious and poorly paid. The big challenge for Albacete in the future will be to create employment to retain the graduates from its university. The generation of more innovative and technologically relevant enterprises and occupations is fundamental to the province's future. Finally, given the rise in the economically active population, the rate of job creation will be a key factor in restraining the level of unemployment. It appears that policies implemented in Albacete over the period 1980-1995 have been appropriate to local needs. This general orientation will need to be maintained in the future, whilst reinforcing those elements which promoted success in the past (improvements to the communications network, cooperation between actors, attracting businesses from outside the region etc.).

References

BBV (Banco Bilbao Vizcaya) (1997) *Renta nacional de España y su distribución provincial (Spanish national income and its provincial distribution)*. Bilbao.

CPLPSDAMS (Consorcio para la promoción socioeconómica del Arco Mediterráneo Sur) (1996) *Albacete, Alicante, Almeria y Murcia. Un espacio común de desarrollo (Albacete, Alicante, Almeria and Murcia, a common development space)*. COCIN de Almeria.

INE (Instituto Nacional de Estadísitica/National Statistics Institute) (several years) *Anuario estadístico de España (Spanish statistical yearbook)*. Annual publication, Madrid.

INE *Encuesta de población activa (Labour force survey)*. Quarterly publication, Madrid.

INE *Censo de población (Census of population).* Several years, Madrid.

INE (1985) *Contabilidad Regional de España 1983.* Madrid.

INE *Padrón de habitantes (Inhabitant survey).* Several years, Madrid.

INE *Encuesta de salarios de la industria y los servicios (Industrial and services wages survey).* Quarterly publication, Madrid.

INE *Movimiento natural de la población (Natural population change).* Annual publication, Madrid.

López Trigal, L. (ed.) (1994) *Zamora. Un espacio de frontera (Zamora. A frontier area).* Fundación Rei Afonso Henriques, Zamora.

OECD (Organization for Economic Cooperation and Development) (1994) *Creating rural indicators for shaping territorial policy.* Paris.

Rosell, J. and Viladomiu, L. (1998a) Creación de empleo no agrario en provincias de baja densidad demográfica. Zamora vs. Albacete. (Non agrarian employment creation in low population density regions. Zamora versus Albacete). *V Congreso de Economía Regional de Castilla y León.* Comuniciones 2, Junta de Castilla y León, Valladolid.

Rosell, J. and Viladomiu, L. (1998b) *Case study leading region in Spain: Albacete.* Universitat Autonoma de Barcelona, Dpto. Economia Aplicada, Barcelona.

Rosell, J. and Viladomiu, L. (1998c) *Case study lagging region in Spain: Zamora.* Universitat Autonoma de Barcelona, Dpto. Economia Aplicada, Barcelona.

Rural France: the Case Study Regions of Les Alpes de Haute Provence and La Nièvre

8

Jean-Pierre Bertrand and Bernard Roux[1]

Introduction

The French leading case study region – Les Alpes de Haute Provence (AHP) – experienced in recent decades an increase in both employment and population. The change in employment was made up of a decline in agricultural employment, a moderate increase in industrial employment and a considerable rise in services employment (Table 8.1). Population growth was both the result of natural growth and inmigration. The lagging case study region – La Nièvre – showed both a decline in employment and population. Although employment in the service sector increased, this increase was insufficient to compensate for the loss of employment in the agricultural and industrial sectors. In this chapter the forces behind the different performance in AHP and Nièvre will be discussed.

The AHP region is mainly rural with a scattered settlement with many small towns and villages. Its administrative town (the *Préfecture*), Digne-les-Bains, is one of the smallest in France. The active agricultural population in the total population is above the national average (Table 8.2). Industry has never developed significantly because of the region's isolation and relatively remote geographical location, far away from regional and national hubs of economic development. The only noteworthy industries were established in the Durance valley, which is the only area easily accessible from other regions. Since World War II, industrial employment (except for construction) has remained fairly stable at around 10% of total employment. The historical turning point in the service sector occurred at the beginning of the *trente glorieuses* (i.e. the three decades of continuous and strong growth after the end of World War II), when the demand for leisure activities in France grew rapidly. AHP entered the dynamics of winter sports holidays with the establishment of several ski resorts. At a later stage – and this is still the case today – the region developed tourism through the enhancement of many natural and cultural resources. So AHP is a good example

[1] With helpful comments from the other members of the French RUREMPLO team: Jacques Blanchet, Jean-Claude Bontron, Åsa Sjöström and Germana Foscala-Baudin.

Table 8.1 Growth of non-agricultural employment and population, 1981-1992 (% p.a.)

	Non-agricultural employment	Population[a]
Alpes de Haute Provence	1.1	1.1
Nièvre	-0.3	-0.3
Most rural regions	0.5	0.3
Intermediate regions	0.8	0.7
Most urban regions	0.3	0.5
France	0.4	0.5

Source: RUREMPLO project.
a) 1980-1992.

Table 8.2 Socio-economic indicators of Alpes de Haute Provence and Nièvre

	AHP	Nièvre	France
Population, 1990	140,300	233,300	56,615,200
Area (km²)	6,900	6,800	544,000
Population density (inh./km²)	20	34	104
Population growth, 1982-1990 (% p.a.)	1.2	-0.3	0.5
Total employment growth, 1981-1992 (% p.a.)	0.5	-0.6	0.06
Non-agricultural employment growth, 1981-1992 (% p.a.)	1.1	-0.3	0.4
Population > 65 years (%)	19	21	14
GDP/inh., 1994 (EU=100)	94	87	116
Sectoral employment share, 1996 (%)			
Agriculture	6	8	5
Manufacturing	12	21	19
Construction	9	6	6
Services	73	68	70
Unemployment rate, 1996 (%)	12	11	12
Population main cities, 1996	Manosque:19,100	Nevers: 58,900	Paris: 2,116,000
	Digne: 16,100	Cosne: 13,200	Marseille: 798,000
	Sisteron: 6,600	Decize: 9,100	Lyon: 445,000

Source: INSEE.

of a region that has made a transition from an economy in which agriculture is the main feature, to an economy in which services are predominant.

In contrast to AHP, Nièvre has a long industrial tradition. The ironworks are part of the history of the Nivernaise industry. Its focal points are the processing of metals and automotive equipment. About 30% of the workers are concentrated in the Loire valley. Traditionally, Nièvre was a region of seasonal emigration of agricultural and forestry workers or long term out-migration towards the Parisian labour market. The tradition of nurses, which prefigure modern institutions of Children Welfare in Nièvre, was among these migrations. Nièvre has experienced a long industrial crisis since the beginning of the 1980s, which continues today. This crisis brings not only closures of enterprises and redundancies, but also a structural change in qualifications and gender of the active population. As a result, industry becomes more specialized and new sectors, like rubber and plastics, electrical and electronics equipment, and paper production, are developing. The service sector in Nièvre consists mainly of activities of a traditional and public type, like trade, transport, health and education. Services for persons, enterprises and tourism are growing.

Socio-economic indicators

AHP and Nièvre are both mountainous regions with valleys, where industrial activities are concentrated, plateaux where agricultural activities dominate along with many rivers and lakes. Although tourism is an important source of employment, a big geographical and climatic difference exists between both regions: AHP has snow and skiing possibilities in winter, while Nièvre suffers from difficulties of access by road during this season (especially in the Morvan mountains). Population density is low in both regions: respectively 20 inhabitants per km² in AHP and 34 in Nièvre (Table 8.2). The urban structure of both regions reflects the low population density and the very modest importance of industry and the functions of its towns, which are essentially commercial and administrative. This bunch of small towns does not constitute a structured network, but a series of small micro-regional centres, serving territories of various sizes, and is rather badly connected to the main regional and national hubs of development (Fig. 8.1). The percentage of old people in both departments is well above the national average. In terms of GDP per inhabitant, AHP shows a better performance than Nièvre, but both are still below the national average.

EU structural policies

In both case study regions the European Structural Funds strongly affect the behaviour of actors. The European Commission does not exactly implement a policy directed at the eradication of unemployment, but its objectives are an effort aimed at the improvement of infrastructure and development, which are influencing the state of the labour market. The main element is that policy makers

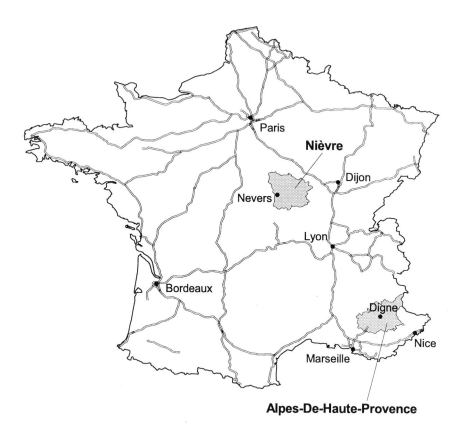

Figure 8.1   Location of AHP and Nièvre in the French context

and other actors are establishing links between European Structural Funds
(ERDF/FEDER, ESF/FSE, EAGGF/FEOGA) on the one hand and the policies
adopted by the French state, the region and the department at the other hand. In
practice no policy is applied in isolation, since each one takes into account the
nature of the other ones. Nièvre is completely covered by the European structural
policies: first, for sustaining the restructuring of the industrial sectors of Cosne-
sur-Loire, Nevers and Decize and second, for consolidating the totality of the
rural areas. In AHP the support was directed principally towards mountainous
agriculture.

Local resources

Poor transport infrastructures in the mountainous areas

Reaching and travelling within AHP have always been problematic. As a peripheral region, AHP has not benefited from national priority funds for the development of roads and railways. The road network is typical for mountainous regions: narrow and winding roads, on which it is impossible to drive fast. This is why transport is time consuming. That is one of the reasons for the concentration of economic activities in the Durance valley, open to the south towards Marseille and wide enough to let a motorway go through. The railway remains a very marginal means of transport.

At first sight, Nièvre has a favourable position: it is close to Paris, there are good railway and road infrastructures, abundant agricultural, forestry and water resources. At a more detailed look, it seems evident that the department remains very isolated, notably because the Morvan mountains are a true obstacle, especially in winter. Consequently, most of the activities have been concentrated in the Loire valley. The isolation of the region is frequently mentioned as an important obstacle to employment development in the region. It is especially true for the rural areas of the central Nivernais and the Morvan massif. On the other hand, the Loire valley and the axis Paris-Nevers are efficiently served, notably by train.

AHP: rural amenities mobilized for development

AHP has a Mediterranean climate, beautiful mountainous landscapes and numerous rivers, which would be torrential if dams did not regulate them. The water that they carry is a reserve for irrigation and a supply of energy and drinking water for the PACA region (Provence Côte d'Azur district). It is quite uncommon in France to consider the air quality as a local resource. Nevertheless, this is the case in AHP, reaching a point where air quality is an advertising point not only aimed at tourists but at entrepreneurs as well. Since the 1960s AHP has started to invest seriously in tourist development. To this end, infrastructure for winter sports and lodgings for tourists have been built. Winter sports holidays have been considered as the main axis of development for mountainous regions during the strong 30-year period of economic growth after World War II. Several ski resorts have been created in areas where everything had still to be developed, ranging from necessary skiing equipment to hotels and other lodgings. Tourism during the summer has also increased considerably. A number of natural parks, like the Luberon and Mercantour, can be found either partly or in their entirety in AHP. The necessity of protecting the environment and the importance of tourism in the region's economy have reinforced the local actors' interest in nature and historical cultural heritage. Like the rest of France, culture has become an integral part of the attractiveness of rural areas. In many places in France, cultural events have been created, which provide a lively atmosphere for the tourist summer season, even in the most remote villages.

The natural and cultural resources have provided the incentive for AHP's transition towards an economy mainly based on tourism. The space available, the mild climate, the beauty of the landscape and the proximity of urban centres on the coast explain the uninterrupted development of secondary homes, whose presence prevents the complete depopulation of certain areas.

Nièvre: trying to reverse the fire wood export model

In Nièvre the most important resources today are its landscapes: mountainous forests in the Morvan and open fields and meadows in the central region. Numerous rivers cross and/or originate in the department, in addition to many canals, lakes and reservoirs created to transport wood by floating. Resources in quick (and mineral) waters are abundant, coming mainly from the Morvan. The department possesses one of the biggest oak forests of France, covering one-third of the surface of the department. During the last centuries wood has been the main resource and the department has supplied Paris with firewood. It is still a resource, but not so important as before. The mines have run out. Finally, the region is endowed with cultural resources (historical, archaeological and religious).

These physical and cultural resources are valorized in various ways in forestry, landscape protection and agriculture. Attempts are made to reverse progressively the historical tendency to export natural resources in their raw state, with little added value. However, in forestry it is very difficult to come out of the model of firewood export towards Paris, which was ruined by the arrival of coal and replaced very partially by the production of Christmas trees or by the selective exploitation of hardwood. The creation of the Regional Natural Park of the Morvan in 1974, in a region with a strong cultural and social identity, was based on the idea that it was a coherent territory and not only an employment basin. A sustainable development strategy has been implemented, based on a human resource base of about 35,000 inhabitants living on 225,000 hectares (15.5 inh./km²) and counting on the important cities at the periphery of the Park like Avallon, Corbigny, Nevers and Autun. The present specialization in agriculture towards the production of lean cattle is based on high quality grazing resources and exploitation structures, which enable enlargement. Agrotourism is promoted by the supply of rural lodges, but often fails because of insufficient local product valorization.

Employment dynamics during 1980-1996

The labour markets in AHP and Nièvre are unbalanced, like in all NUTS2 regions of France and in the national territory as a whole. This imbalance results in a high level of unemployment: 12% of the AHP's active population, 11% in Nièvre and also 12% in France as a whole (1996). The reasons for this high level of unemployment are numerous and complex. We can mention the effects of the globalization of the economy as well as the insufficient flexibility of employment regulations and the labour cost, even if this last factor is not so evident in France compared with other European countries. These general factors exert their influence on territories such as the departments of the AHP and Nièvre. Wage levels, institutions that contribute to the functioning of the labour market and migration of workers are the main local factors influencing employment. The average wage was respectively 97% (AHP) and 92.5% (Nièvre) of the French level in 1995. This difference is not sufficient alone to encourage companies to relocate. The quality of infrastructure, the qualifications of the labour force, the regional, national and European policies are competitive factors, which are much more important for the attractiveness of a region. Besides, to understand the dynamics of the labour markets some structural constraints should be taken into account, like the imbalance between the industrial and service sector, the dependency of the largest industries on decision centres outside the region, the insufficiency of public transport between cities and rural areas, the importance of migration flows, the imbalance due to the seasonal nature of some activities and the competition of 'black market' jobs.

Employment development

In the study period employment in agriculture decreased in both regions, whereas that in services increased. Employment in industry in AHP more or less stabilized, but that in Nièvre experienced a decline (Fig. 8.2). These employment dynamics resulted in an increase in total employment of 0.5% per year in AHP in the period 1981-1992, while Nièvre suffered an employment loss of 0.6% per year.

Industry has always been a small sector in AHP. During the study period chemical industries and household equipment industries showed large employment losses, which were more or less compensated for by new jobs in the agrifood industry and electrical and electronic engineering. In Nièvre industrial strategies are rather defensive in the big basic sectors of metallurgy, chemical and wood industry, because of the restructuring process, which provokes closures and massive redundancies. We can add that the food industry is rather weak compared with the agricultural production.

In both regions public services contributed substantially to employment creation. In Nièvre many jobs were created in the branch of services to persons and enterprises. In this branch an important network of subcontracting firms (more than 180 enterprises) is active, working mainly for the automotive sector in very pointed market gaps: wiring, computer services, treatment of surfaces, plastic and rubber components. This network employs a lot of qualified workers. The big commercial centres are also growing, often at the expense of the small rural retailers. Finally, public employment, notably in the sector of health itself, also develops quickly.

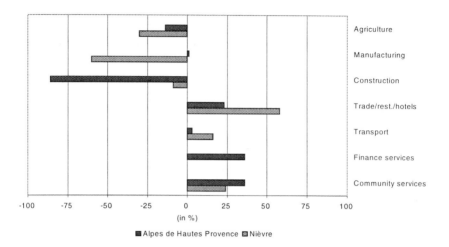

Figure 8.2   Share of branch in gross increase/decrease in employment in Alpes
de Haute Provence and Nièvre, 1980-92[a](%)
Source: INSEE.
a) For AHP: 30,700 employed persons in 1980 and 35,800 employed persons in
1992; gross increase was 6,600 and gross decrease was 1,500, resulting in a net
increase in employment of 5,100 employed persons. For Nièvre: 88,000
employment persons in 1981 and 82,000 employed persons in 1992; gross
increase was 7,000 and gross decrease was 13,000, resulting in a net decrease of
6,000 employed persons. The bar for trade, restaurants and hotels for Nièvre
includes employment changes in the financial services.

Actors

Strategies of policy makers

There are many people's representatives who play a role in local development
policy-making in France: deputies and senators, district and regional councillors
(elected by the department and the region), mayors and municipal councillors.
Each one defends his/her own interests and his/her own political beliefs and, at
the same time, each one takes into consideration, the interests of the district's
territory which he/she brings to the attention of the highest levels of the political
sphere, namely the State and the European Union. Policy makers in AHP are able
to identify needs in the region and to deliver policies according these needs. They
also have good contacts with the upper level authorities. In the 1980s Nièvre had
several powerful politicians, like François Mitterand, who was councillor and
Mayor of Château-Chinon, who went to national politics. These policy makers
favoured the public sector. Today, policy makers in Nièvre are less powerful and
priority is given to employment development in the private sector in the regional
context. Local policy makers do not have good access to the regional policy

makers in Dijon. On the whole, their attitude is more defensive than forward-looking.

Both regions applied a policy favouring the concentration of industry in the valleys of La Durance and La Loire, which involved the construction of highways, railways and other infrastructures linked to companies. In AHP, financial support was given to ski resorts, in order to relieve the difficulties with the accommodation capacity, the accessibility and attractiveness of the sites. Further, in order to maintain a minimum level of economic activities in the thinly populated rural areas of AHP and hence to prevent depopulation, economic assistance has been given to agriculture and services in those parts. The quality of the air and water in AHP is an argument for the settlement of retirees and the location of health services. Municipalities in AHP are helping the settlement of enterprises, traders and artisans. Policy makers in Nièvre were also concerned with the prevention of the loss of economic functions in small towns (shops and public services). So regional and local authorities tried to use national and European employment policies for maintaining basic rural activities and services by supporting local trade and public services, green tourism and other new economic activities. Next, they gave support to open-air activities like walking, water sports, etc.

Industrial strategies

The strategies concerning the setting up of companies in AHP in the 1960s were dictated to a large extent by the nuclear plant in Cadarache (5,000 jobs), located a short distance away in the neighbouring department. The idea was to meet the need for subcontracting and to benefit from the repercussions of high-tech. Hence Elf-Atochem and Sanofi and some other sub-contractors have settled in the Durance valley and structured its economy. These establishments emphasize the territorial duality of AHP: on the one hand there is a densely populated Durance-Bléone Valley (130 inhabitants/km^2) with several industries on only 7.5% of the territory, and on the other hand there is a large area (92% of AHP and 10 inhabitants/km^2) with mainly agricultural and tourist activities. This duality explains the strategies of consolidating activities and employment in the Durance-Bléone Valley, where they have reached a significant level, and resisting depopulation elsewhere. Entrepreneurs (both local and in particular newcomers) are innovative in searching for new initiatives and products as well as new ways of marketing of products. The technical and managerial knowledge of entrepreneurs is less developed.

Due to the long industrial tradition with large multinational firms, entrepreneurship in Nièvre is less developed than in some other French regions and entrepreneurs are risk averse. The innovating capacity is quite weak, except in the automotive sector. In the agro-industry and in the processing of wood, the follow-up strategies are on the whole rather defensive: entrepreneurs usually exploit the first steps of the production process, or even trade in raw materials (rough timber, cereals, oilseeds and lean cattle). This limits the value added on site and the different skills necessary in more complex processes of processing. Companies reduced their workforce during the recent years, notably in the cooperative sector. Some big enterprises in rural areas suffer from high transport costs. Multinationals in both regions do not take local considerations into account and their employment strategies are based on decisions taken outside the region. The strategies of the small and medium sized enterprises are individual, although they often share know-how.

Multifaced agricultural sector

Agriculture in both regions covers a wide range of products. In AHP this variety results from the diversity in natural conditions. Mediterranean crops such as fruit trees, olive trees and lavender can be found in the department, as well as the rearing of dairy herds in the mountains. Farm holdings are mainly specialized family farms. The farming system is more determined by the farm's geographical location than by its size. Three technical and economic farming systems are predominant:

1. cereal and large-scale farming (25% of holdings in 1995), particularly durum wheat and typical production of the Provence lavender and lavandin, which are facing a serious crisis, given the decline in cultivated area and the competition of synthetic essential oils;
2. extensive sheep and goat rearing (24% of holdings), a farming system typical of mountain areas and dry hills, enabling the development of low productivity land;
3. fruit growing (18% of holdings), mainly producing apples and concentrated in the irrigated part of the Durance valley.

It should be stressed that the olive tree, a symbol of Mediterranean production, which was extensively grown in the south and west of the department in the past, has found its dynamism again as the result of the growing demand for olive oil. At the department's level, these structural characteristics are at the root of reduced land and labour productivity and, finally of the small farming revenue. Over the period 1993-1996, the gross revenue per farm holding in AHP was one-third below the national average. On average, the agricultural sector is not very competitive, which explains the constant fall in the number of farm holdings and employment.

The situation of agricultural employment in the Durance valley and in some other small areas in AHP is totally different from that found on the plateau and in the mountainous areas. Here farmers look for workers outside the farm, i.e. on the job market, which are needed in the orchards and the intensively irrigated vegetable production. The peaks of labour demand are in summer, at the beginning of the autumn (harvest) and in winter (tree pruning). The labour demand is never totally met by the local workforce and this explains the flow of workers coming from other French regions and from abroad, mainly Spain and Portugal. According to a study on fruit farming carried out in the north of Sisteron, on an annual basis local workers account for 60% of the working hours of seasonal workers, workers from other regions account for 13% and foreigners for 27% (Baldassari, 1992). However, at harvest time, there is about one temporary worker per hectare and the share of foreigners increases to 40%.

Agriculture in Nièvre comprises at least three regions: the Morvan (production of wood and Charolais cattle breeding), the Loire valley (Pouilly wine production) and the Nivernais central region (Charolais cattle breeding). Grain production is predominant on a strip, that goes from Cosne-sur-Loire to Clamecy. This type of agriculture is capital intensive and employs few people per hectare. The production of lean beef cattle occurs either in specialized systems of extensive breeding (suckler cows) or in mixed farming-breeding systems. The extensive character of this production provides few opportunities of job creation in this activity. In spite of efforts made by some professional organizations, the fattening is practised at a marginal level and it often concerns only some of the livestock.

In Pouilly-sur-Loire, a valley along the Loire, a small well-known vineyard, classified in AOC, produces the *Pouilly Fumé* wine and in a smaller amount the *Pouilly-sur-Loire* wine. Being an old region of table grape production for the Parisian region, the vineyard has been replanted with *Sauvignon* grapes since the 1960s (about 1,000 hectares currently) for the production of the *Pouilly Fumé,* while 40 hectares were maintained in *Chasselas* for the production of the *Pouilly-sur-Loire.* This wine producing valley is one of the few examples of agricultural branches that have created jobs during recent years. The valley is influenced completely by the vineyard, as much in terms of employment, as in terms of induced purchases: vats of steel, oak barrels, input purchases for treatments of the grapevine. At the level of employment, the work of the grapevine presupposes the enrolment of one worker per 3-4 hectares. As vineyards have an average size of between 8-12 hectares, about 300 people are employed full-time in the grapevines. To this, it is necessary to add about 60 seasonal workers for dressing the vine. The cooperative employs eight persons for wine producing, handling and merchandising. The cooperative cellar has a capacity of 7,000 to 10,000 bottles per day, for a turnover of 25 millions of French francs, of which 55-60% is exported. Finally, the purchasing power of winegrowers nourishes local trade. This region makes therefore an exception in relation to tendencies of decreasing agricultural employment observed in the other rural areas of Nièvre.

Farmers' strategies: enlargement, diversification, high-quality production

The main farmers' strategy in both regions is to increase the size of farms and productivity. With a restricted land area, such a strategy in a strong competitive environment tends to limit employment, notably in extensive farming systems. It has positive repercussions on country planning. First, it could contribute to landscape conservation. Second, it helps retaining a minimum number of inhabitants in less populated areas (even if the absolute number continue to decrease). Finally, it ensures the continuity of a permanent economic activity in disadvantaged areas where other sectors are missing (industry) or are of a seasonal nature (tourism). The creation of Natural Parks reinforces these effects.

Others farmers' strategies refer to pluriactivity, diversification of activities and the production of high quality products. In AHP, 19% of the farm managers and many more of the spouses were engaged in off-farm activities in 1995. However, pluriactivity is not common in the thinly populated areas with few non-agricultural jobs. In the Durance Valley pluriactive farmers are employed in industries. Agricultural production on the AHP's plateau and in the hilly and mountainous areas is positively influenced by its natural environment, which allows farmers to adopt a quality-oriented strategy. In AHP, young farmers tend to introduce new crops and to adopt new farming systems. Soft fruit, herb growing and organic farming are amongst the alternatives most frequently explored. Diversification through cattle rearing is also present, such as free-range chickens, goats and beehives. Some well-known traditional products, such as perfume plants (lavender, sage and mint), Sisteron lamb (230,000 animals per year), Banon goat cheese, quality lavender honey, Riez and Banon truffles and olive oil from Haute Provence, to be awarded with the AOC quality label, are renowned thanks to a constant demand.

In Nièvre some examples of diversification could also be mentioned. A few new original orientations appear in the department. For instance there is a cereal producer, who is involved in the production of dried flowers. This activity mobilizes the family workforce and seasonal workers. Outlets are found at a neighbouring enterprise that manufactures some floral arrangements (49 employees), by sale on the local markets and by direct sale on the farm. This producer considers other diversification: the planting of walnut trees on two hectares and the creation of a 'rural lodge'. In the Morvan new activities concern the cultivation of Christmas trees, of which Nièvre is currently the largest French producer. This activity is part of traditional forest activities, but does not always go well with other forestry activities. Other more specific markets, but well adapted to the natural habitat of the Morvan, exist as well. Several producers are involved in medicinal plant production. To summarize, we can say that here most of the actors give priority to the family workforce and minimize salaried workers' employment.

Diversification strategy towards agrotourism

Agrotourism is a promising prospect for farmers, as the demand for rural tourism is growing. However, many factors hamper farmers' participation in tourist activities, as they must have capital to build lodgings, have sufficient workforce to take care of tourists and clean the accommodation, and produce the raw material for the preparation of meals for guests. Due to these difficulties it is estimated that only about 5% of farm holders in AHP are involved in tourist activities. The most common way is that they provide *gîtes* or guestrooms under the *Gîtes de France* quality label. In rural areas of Nièvre agrotourism is still a marginal activity for farmers: there are 18 *gîtes* (440 rooms), two holiday-villages (500 rooms) and 27 chalets (170 rooms). Together with the 24,000 secondary residences, this accommodation can welcome close to 120,000 people. Local policy makers are promoting agrotourism among farmers, as farmers have insufficient knowledge of this alternative.

Networks in AHP stronger than those in Nièvre

Internal networks function well in AHP. Nevertheless, a certain amount of competition remains between towns, subregions and tourism resorts. Local administrative layers are able to identify the needs of the region, for example the stagnation in the quality of the tourist accommodation, and try to relieve these problems. On the contrary, the intervention of internal networks in employment development in Nièvre, like the Chamber of Trade and Industry, the Chamber of Agriculture, Worker Unions, Farmers Syndicates and Enterprise Associations, is relatively weak, except for the Workers Unions for which employment protection is a priority. In Nièvre it is traditionally the role of the State to manage the conflicts and competition between the actors, and to define a long-term strategy of employment development.

In AHP external networks are used to get financial support from the national and EU funds for regional development. Secondly, external networks are used for lobbying for the interests of the tourist sector in the mountainous areas, for example a national association of the mayors of municipalities located in the mountains and the skiing resorts, who cooperate with those in other Alpine regions. Thirdly, there are external relations with multinational firms, in order to protect employment in the subsidiary business in the region. On the whole, external networks have a positive impact on economic activities in AHP and therefore also on employment, mainly in the tourist sector. Companies from abroad have moved into AHP thanks to successful promotion. In Nièvre external networks take two main forms: multinational or big national firms networks and political networks. Nièvre has some networks of the first type with SNCF (the national railway company, which has important facilities for repairing of trains in Nevers) and the private group Peugeot. In political networks, the Socialist Party has the most important influence. The relations with the regional authorities are weak, except for some regional infrastructure building. These external networks contributed to the maintenance of employment (some industries, especially automotive and rubber and plastics manufacture) or to increase employment (especially in the tourist and public sectors, health, education and social services).

Endogenous or exogenous forces as engine behind employment growth

Employment growth is a mix of endogenous and exogenous forces in both regions. In AHP endogenous forces usually played the initiating role in this mix. Especially, local administrative layers play an important role in identifying the needs of the region, put them forward to upper level authorities and ask for funds. In the tourist sector there is cooperation among actors, which have the capacity to improve their situation. The two biggest companies in the region, Elf-Aquitaine and Sanofi, belong to multinational firms. In Nièvre the initiating role comes from exogenous forces, due to the ageing of population, the lack of entrepreneurship, education and innovative capacity of local actors. Nevertheless, the department has some strengths: some good road and railway infrastructure and some natural and cultural resources. External firms could use all these favourable factors. All local institutions are waiting for that.

Key factors of success and decline in employment dynamics

At the beginning of the 1950s, few people would have doubted the poor economic perspectives of AHP. It was a poor mountainous and agricultural region. On the other hand, Nièvre, with its natural resources (wood, mines and agricultural products) and the proximity of the Parisian markets, had a lot of good cards to play. As we have seen, AHP has achieved a relative success in employment creation, while employment in Nièvre has declined. The key factors behind this difference between the two regions are discussed below.

Key factors of success in AHP

(1) Attractive landscape

The beautiful diversified landscape associated with a Mediterranean climate (modified by altitude) and the possibilities of winter skiing are elements, which attract workers, artisans, entrepreneurs, tourists and retirees. All these actors need services, which create jobs opportunities.

(2) Positive migration balance

AHP has a positive migration balance, essentially made up of highly educated people, setting up enterprises, unemployed people, looking for a cheap place to live and retired people. There are many seasonal labourers working in tourism and agriculture (fruit harvesting). This positive migration balance results, together with the attractive landscape, in a positive atmosphere in AHP.

(3) Active attitude of actors

The positive atmosphere in AHP brings about an active attitude of actors, who face and tackle innumerable challenges in the field of natural conditions, infrastructures, access, diversification of activities, attractive tourist and industrial sites and the maintenance of services, like shops, public services and local trade in small rural towns. Policy makers are able to identify the needs in the region and to deliver policies according to these needs. They have also good contacts with the upper level authorities. Entrepreneurs (both local and in particular newcomers) are innovative in searching for new initiatives and products as well as new ways of marketing of products. Combined efforts by policy makers and entrepreneurs towards an upgrading of the tourist accommodation resulted in a reversal of the stagnation in the tourist sector.

Key factors of success and decline in Nièvre

(1) Attractive local resources

Nièvre has natural and cultural resources, like a beautiful diversified landscape, rivers, lakes and the Morvan mountains, which are attractive for summer activities. These rural amenities are the base for tourist activities, but these are less intensively exploited than in AHP.

(2) Negative migration balance

Nièvre has a negative migration balance, made up of emigration of young people under 30 years and an immigration of retirees. Emigration has been a strategy for workers for a long time and consists of both seasonal migration of agricultural and forestry labourers and long-term migration to the Ile de France region as civil servants in the postal, electricity and railway transport services. Usually, these long-term emigrants come back into the region after retirement. In this sense, the proximity of Paris with higher wages has played a negative role in the local labour market. The continuous emigration of young workers erodes the human resource base in the region and affects the atmosphere negatively.

(3) Passive attitude of actors

The attitude of the actors can be characterized as 'rather passive'. This is reflected in the tradition of outmigration: people tend to look for employment opportunities as civil servants outside the region rather than trying to exploit local sources and create their own employment opportunities. The relatively large number of big multinational enterprises, which can be perceived as signs of 'development from outside', heavily promoted by intervention of the national authorities, also emphasizes this passive attitude. The implications of this attitude emerged to full extent after the collapse of the large industrial firms in the 1980s: the restructuring of the regional economy was hampered by a lack of local entrepreneurship, a defensive attitude of policy makers and a restricted range of labour skills. Most entrepreneurs are risk averse and not very innovative: they prefer to exploit the first steps in the production process and thereby miss many chances for value added. Policy makers tend to be defensive and less forward-looking and not able to establish good working contacts with the regional layers in Dijon. Nevertheless, there are some local actors, who are rather dynamic and establish market niches, like the subcontract firms network, the farmer with dried flowers and the wine producing network in the Loire valley. The impact of this dynamic minority is, however, too small to reverse the declining trend in employment.

Concluding remarks

A main difference between AHP and Nièvre is the attitude of local actors: in AHP actors tend to be active and directed towards employment creation while actors in Nièvre tend to be passive and wait for jobs established by external actors. The active attitude of actors in AHP enhanced employment growth, and such an attitude would have been helpful in Nièvre, to face the employment decline due to the industrial crisis in the 1980s.

A large number of French rural regions are faced with a situation in which more jobs are created in the urban parts than in the rural parts. But paradoxically unemployment is frequently less severe in the rural parts: unemployed people do not stay in the small rural villages, but leave in order to try their chances in the city. These heavy tendencies are practically impossible to thwart. Policy makers must recognize this reality, accompany and sustain the restructuring process with appropriate measures and take care of its people.

In its less populated areas, AHP is subject to difficulties resulting from a lack of sufficient services. Actors have designed a strategy in order to fight against depopulation. Firstly, the State has kept administrative jobs in health, education, and administration, and grouped services in larger villages. Local authorities and small entrepreneurs have set up specific actions in order to support the functioning of the local economy, in terms of supply of provisions (small shops, offering several services) and marketing (sale cooperatives), sometimes supported by European funds. The combination of national and local action in AHP shows that there is a possibility of slowing down the process of depopulation of the countryside and to obtain a population increase in those sparsely populated areas.

With regard to the repopulation of the sparsely populated countryside in AHP attention should be paid to the migrant population. There are no demographic studies on this subject, but the available data suggest that there are two categories of migrants: those who look for a better quality of life and settle down by starting their own business, mostly artisans, and those who move to more distant rural areas because life is cheaper and who are often unemployed or receiving a RMI (*Revenue Minimum d'Insertion* i.e. unemployed receiving a minimum wage). For the latter, this choice is double-edged since settling down far away from urban areas without a job, reduces the possibilities of finding a job. Therefore, migration of people to small isolated villages does not always result in an economic revival of marginal areas. Nevertheless, the local authorities welcome the arrival of the unemployed, since they can then justify the demand of public funds for the local organizations.

The agricultural sector also provides some opportunities to prevent depopulation by shifting from more extensive ways of production to more intensive forms like fattening of livestock and meat processing. The search for higher value added, like labelling (AOC) and/or providing high-quality products, would support the creation of employment. In the same way, small market niches (production of medicinal plants, cheeses and dried flowers) can create jobs if they are in line with strengths of the sector and opportunities in the market. Finally, the large number of secondary homes (32,000 in AHP and 120,000 in Nièvre) contribute to a certain population base in rural areas, although only temporary.

The public authorities of the territorial communities of both departments are encouraging employment. Mayors turn into real chiefs of enterprise orchestrating, trying to attract enterprises and to define a coherent strategy of development. In addition, national employment policies also have an important impact on employment in rural areas, such as recent policy measures towards the creation of new jobs for young people in the education sector and the 35-hour week. Such measures can modify the competitive position of sectors and regions.

References and further reading

Aubert, F. (1996) *Marché du travail en milieu rural; Contribution à l'analyse et application à la Bresse Bourguignonne (Labour market in rural areas; Contribution to the analysis of the Bresse Bourguignonne case study.* Thèse de Doctorat, Université de Bourgogne (Thesis of the Burgundy University), Dijon.

Agence Nationale pour l'Emploi (ANPE) and Région Provence Côte d'Azur (PACA) (National Agency for employment and Provence Côte d'Azur district) (1996) *Rapport d'activité (Activity report).* Marseille.

Baldassari, D.M. (1992) *Les arboriculteurs du nord sisteronnais confrontés au problème du recrutement de leur main d'œuvre (The North Sisteron fruit producers are facing recruitment problems of their manpower).* Mission locale pour l'insertion sociale et professionnelle des jeunes (Local delegation for social and professional insertion of young people), Château-Arnoux.

Bertrand, J.P. (1998) *Agriculture and employment in rural regions of the EU; Case study: the département of Nièvre, France.* INRA, Ivry sur Seine.

Comité départemental du tourisme et des loisirs des Alpes de Haute Provence (Tourism and leisure departmental committee of AHP) (1994) *Le tourisme en chiffres dans les Alpes de Haute Provence (Figures about tourism in AHP).* Digne-les-Bains.

Conseil Régional PACA (Regional Council PACA) (1994a) *Dossiers 'parlons emploi' 8/2; L'emploi en PACA et 8/5; Les pistes pour l'emploi (Employment in PACA district and ideas for employment strategies).* Marseille.

Conseil Régional PACA (1994b) *Document unique de programmation du plan de développement des zones rurales - objectif 5b (Single programming document for rural development - objective 5b).* Marseille.

INSEE (Institut National des Statistiques et des Etudes économiques) (1993) *Portrait des zones d'emploi de la région PACA (Portrait of employment areas of the PACA district).* Marseille.

INSEE (1996a) *Emploi et chômage de 1990 à 1994: évolutions différentes selon les zones (Employment and unemployment from 1990 to 1994: differentiated evolutions in areas).* INSEE Bourgogne Dimensions, no.32, March.

INSEE (1996b) *1946-1996: 50 années de transformations économiques et sociales en Bourgogne (1946-1996: 50 years of economic and social transformations in Burgundy).* INSEE Bourgogne Dimensions, no. 33, April.

INSEE and Préfecture de la région Bourgogne-Conseil régional de Bourgogne (1996) *Atlas de Bourgogne (Atlas of Burgundy).* Les dossiers de Dimensions, INSEE no.15, June, Bourgogne.

Institut National de la Recherche Agronomique (National Institute of Agronomic Research) INRA-INSEE (1998) *Les campagnes et leurs villes (Rural areas and their cities).* INSEE, Paris.

Préfecture de la région PACA (PACA district Prefecture) (1997) *Evaluation intermédiaire du programme Européen (1994-1999) (Final report of the intermediate evaluation of the European programme (1994-1999)).* Marseille.

Roux, B. and Foscale-Baudin, G. (1999) *Case study: the department of Alpes de Haute Provence, France.* INRA, Unité d'Economie et Sociologie Rurales, Grignon.

Comparison of the Italian Case Studies: Pesaro and Macerata

9

Roberto Esposti and Franco Sotte

Introduction

The case study regions in Italy belong to the part of the country that has been called 'third Italy'. The regions of third Italy clearly differ from the older industrial regions of Northern-Western Italy based upon large sized firms and from lagging Southern regions supported by public funds. In the last decades the third Italy experienced an intense spontaneous industrialization process characterized by small family firms often concentrated in highly specialized areas (industrial districts) (Fuà, 1988; Becattini, 1998).

The Marche (NUTS2) region is the southern part of the third Italy. Both Pesaro and Macerata are provinces of the Marche region and are typical examples of its industrial development. They have a similar geographic and socio-economic character and consequently it is easier to identify the main reasons behind the differing employment performance in the last decade. Within the period 1982-1995 the growth rate of non-agricultural employment in Pesaro is +0.1% per annum while it is -0.4% for Italy. For this reason Pesaro is a legitimate leading region. The growth rate of non-agricultural employment is -1.0% per annum in Macerata and therefore Macerata is a legitimate lagging region. Table 9.1 reports some general employment dynamics of both provinces. Looking at employment growth, both show a decline in total employment, although it is larger in Macerata. The difference between the two provinces emerges when only non-agricultural employment dynamics are considered; in this case Pesaro shows a positive behaviour.

The two regions are quite similar in terms of population, land area and density, although Pesaro is a little larger. They share also similar demographic development with a very low increase, in line with the national trend. The unemployment rate is low. The distribution of employment between sectors, shows the important role played by industry especially in Pesaro; the services share is about 50% in both cases while agricultural employment plays a greater role in Macerata. Although classified as intermediate rural region, Pesaro is an example of a region in which the agricultural sector plays a marginal role, at least

Table 9.1   Annual average employment growth, 1982-1995 (%)

	Total employment	Non-agricultural employment
Pesaro	-0.5	0.1
Macerata	-1.8	-1.0
Italy		
Total	-1.2	-0.4
Most rural regions	-1.3	-0.4
Intermediate regions	-1.2	-0.3
Most urban regions	-1.1	-0.5

Table 9.2 General information about the case study regions

	Pesaro	Macerata
Population, 1995 (1000)	337	297
Size (km^2)	2,890	2,770
Population density (inh./ km^2)	120	107
Population growth, 1980-1995 (% p.a.)	0.11	0.15
Population > 64 years, 1991 (%)	18	16
Sectoral employment structure, 1995 (%)		
Agriculture	5	10
Industries	43	39
Services	52	51
Unemployment rate (%)		
1985	7.3	3.7
1995	4.9	5.5
GDP per capita, 1990 (ECUs 1990)[a]	14,050	14,600
Main cities (inh.)	Pesaro: 80,500 Fano: 48,700	Macerata: 43,000 Civitanova: 37,300

Source: ISTAT.
a) GDP per capita in Italy is 14,360 ECU and in the EU 13,360 ECU.

from an employment point of view (Table 9.2). In terms of GDP per capita, both provinces are above the EU average; however, in Macerata it is a little bit higher than the national average while it is below for Pesaro.

Local resources

The Marche region with its various industrial districts is one of the most interesting examples of the North-Eastern-Centre (NEC) Italian rural industrialization process. As a rural region of about 1.4 million inhabitants and 960 thousand hectares, Marche lies in central Italy, bordering to the west on the

Apennine mountains and descending to the Adriatic coast to the east. It has no large urban centres: there are no towns with more than 100,000 inhabitants and there are only four towns with more than 50,000 inhabitants. It is, therefore, a region without an apparent centre-periphery hierarchy. The recent industrial growth in the region has been based on a highly localized and specialized industrial concentration of traditional manufacturing (mostly clothing, textiles, footwear, furniture, but also machinery).

We focus in particular on two provinces in the Marche region: Pesaro and Macerata. Both of them well represent the regional development pattern. They are characterized by relatively small urban centres, by a marked industrial concentration and constantly able to adapt to changing global markets, this being the case of the furniture district in Pesaro and of the footwear district in Macerata (Fig. 9.1). However, outside these successful areas both provinces display a diversified rural system of areas undergoing industrial decline and marginal areas, especially mountainous ones. Therefore, these two provinces are interesting examples of different shapes of rurality within the same region. Indeed, the different forms of rurality are functionally related, in that they are parts of the same development process (Esposti and Sotte, 1998a, b, c).

According to the OECD definition (OECD, 1994; 1996) in 1991 the two provinces were classified as significantly rural (or: intermediate rural) while applying the definition in 1951 we get that both are predominantly rural. During

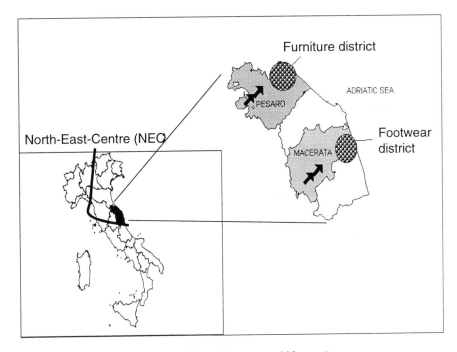

Figure 9.1   Geographical position of Pesaro and Macerata

this period a very intense industrial growth took place. However, this shift in the degree of rurality was not caused by population growth; rather, it was due to a large-scale redistribution of the population across the territory. Population increased greatly between 1861 and 1951 and remained quite stable between 1951 and 1991 in both provinces. However, if we consider the density at the municipality level, on which the OECD definition is based, it exhibits the opposite behaviour: it remains quite stable between 1861 and 1951, while its heterogeneity greatly increases between 1951 and 1991. This means that almost the same population was consistently redistributed across the province's territory, concentrating in some urban centres along the Adriatic sea, the zone best endowed with infrastructures (the motorway and the north-south railway are close to the coast) (Figs 9.2 and 9.3).

During the period of most intense industrial growth (1951-1991), there is no significant population growth; rather, it shifts to centres corresponding to the main industrial concentration constituted by the local industrial districts. The same shift is apparent in infrastructure, public utilities and facilities, and so on. Besides this shift, one notes the change of the provinces in terms of rurality from predominantly rural to significantly rural. This concentration process is closely and evolutionarily linked to industrialization. Services tend also to concentrate where most of population has congregated, providing all the tertiary activities linked to urban centre economies. Manufacturing is mainly concentrated around the main urban centres: the connection between the industrial district and the urban concentration of population and services emerges clearly. This connection is functional to the needs of the industry clustering, when it reaches its critical mass, attracting labour force, activities, services and infrastructure and causing cumulative reinforcement of the concentration itself.

Consequently, the inner and mountainous parts of the two provinces experience progressively an increasing marginalization: decline of population and of human capital, low infrastructure endowment and a low provision of services. Nevertheless, some industrial areas came into being, although increasingly fragile and affected by sectoral crisis. These parts of the provinces have been selected as Objective 5b areas; in the case of Pesaro, there has also been an inner part involved in Objective 2 programmes from the late 1980s due to the crisis of the local textile sector.

Although these geographical characters and the territorial evolution of population and activities are extremely relevant in the long-run development of both cases, they play nevertheless a very poor role in explaining the different recent employment dynamics. Comparing the two provinces, local resource endowment and distribution follows the same pattern and the difference in their employment growth is affected by other complex reasons, which will be explained in next sections.

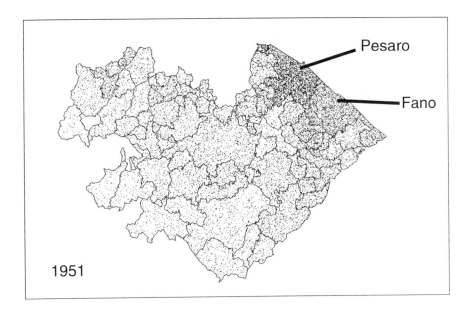

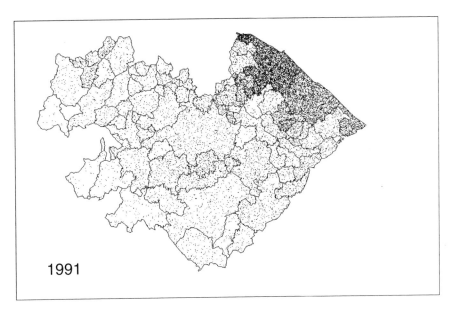

Figure 9.2 Distribution of population and main centres in Pesaro, 1951 and 1991(1 dot = 20 inhabitants)

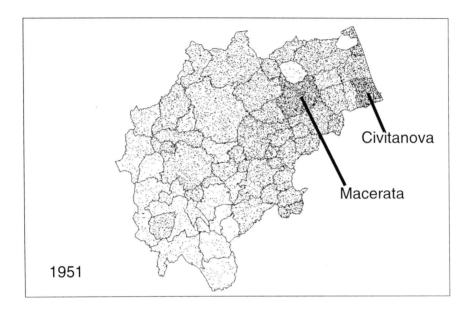

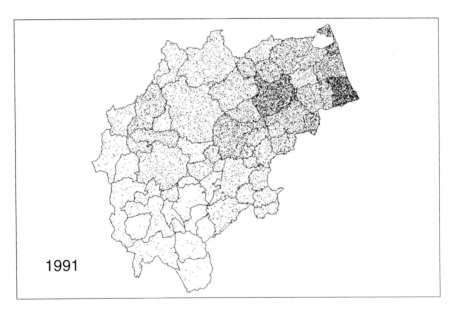

Figure 9.3 Distribution of population and main centres in Macerata, 1951 and 1991 (1 dot = 20 inhabitants)

Economic activities

To understand the reasons of the different employment patterns it is useful to emphasize the main similarities between the two regions in terms of economic development, so that the differential key factors can better emerge. The economic growth of both regions is considered quite recent. A great step forward has been possible due to the intense industrialization of the region that started from the late 1950s. There are two main characteristics of the industrialization process of the region: high specialization in few branches and high territorial concentration. Many small and medium sized enterprises concentrated in limited areas determined by high specialization and external economies. These features define what is usually called the 'industrial district'.

The main and traditional specialities of the regions are wood furniture in Pesaro and footwear in Macerata. These specialities are concentrated in particular in the coastal areas of the regions where most people, firms, activities and infrastructure have recently cumulated. On the base of this intense industrialization three main aspects can be found in both cases:

(1) Preconditions (before the 1980s)

- *Social structure:* a rural family structure, which allowed high flexibility, pluriactivity and low-labour cost; moreover informal knowledge and skills on low-tech and craft processes were embedded in the rural tradition. Both factors produced many industrial initiatives.
- *Market opportunities:* the rapid growth of demand, particularly in the textile-clothing and furniture sector, created many market opportunities both at national and international level.
- *External economies:* the many small family firms concentrated in specialized areas were creating positive externalities such as an efficient local market, diffusion of knowledge and entrepreneurial spirit.

(2) Structural decrease in agricultural employment (1980-1995)

This is a typical effect of rapid industrial growth (Henry and Drabenstott, 1996; Bernat, 1997). Industry attracts labour providing higher wages and better job opportunities. Macerata and Pesaro were typical and traditional rural regions until the 1960s with most of the labour force employed in agriculture. Therefore the reduction of employment in the sector was expected. The phenomenon has been more intense in Pesaro, due to less fertile and rich agricultural resources and to the intense industrial growth in the last decade that has been more successful than in Macerata. In any case the structural trend is the same in both provinces and a further decrease in agricultural employment is expected.

(3) Temporary stagnation of employment in services (1980-1995)

A reduction, or a very weak increase in employment in services is probably a common result in many Italian provinces. Evolutionary patterns of the service sector are a clear reaction to the increasing need of a rational use of resources and of flexibility in facing market opportunities and global competition. Although it produced a reduction of employment, this recent process can therefore be a source of future opportunities. The rationalizing process in the various types of services was as follows:

- *Tourist services:* the increasing competition with other European countries forced the local coastal tourist system to reorganize towards larger sized firms and higher quality services. Agrotourism, cultural and artistic tourism are becoming important employment opportunities in the inner areas.
- *Wholesale and retail trade:* the traditional family structure of these services is becoming progressively marginal, being substituted, at least in the coastal area, by more rational and larger activities.
- *Public sector:* the public budget crisis, which became dramatic in the early 1990s, forced a great rationalization in public services. In many cases, previous employment was excessive with respect to the quantity and quality of the services. Local and national policy makers were then forced to start a rationalizing process. This rationalization had two different effects: a more efficient use of the available labour force along with a reduction in hiring young labour force and a tendency to externalize many activities toward private services. However, this tendency is probably temporary as the rationalization of the public sector is coming to an end.
- *New private community services:* the reducing role of the public sector and services allowed the creation of new private activities providing several kinds of services assuring higher quality, higher efficiency and pursuing social aims. Employment perspectives are particularly interesting because of the great number of new services and activities, which these firms (especially cooperatives) are able to provide.

What clearly emerges from Table 9.3 is that neither agricultural nor service employment explains the difference between Pesaro and Macerata in terms of employment creation within the period 1982-1995 but that the development of

Table 9.3 Employment evolution in the two provinces, 1985-1995 (in thousands)

	Pesaro		Macerata		Total growth rate (%)		
	1985	1995	1985	1995	Pesaro	Macerata	Italy
Total employment	139	132	138	115	-5	-17	-13
Agriculture	17	7	19	11	-58	-42	-42
Industry	45	56	57	46	24	-19	-13
Services	77	69	62	58	-10	-6	-8
Unemployment	10	7	5	7	-35	25	25

Source: ISTAT.

industries makes the difference. However, this does not imply that Macerata is undergoing an industrial decline compared with Pesaro. Although the rural industrialization process has reached a mature stage in both provinces, generating a complex industrial-urban system, this is not to imply that it has achieved long-term stability and sustainability (Saraceno, 1995). On the contrary, maturity means that the previous stage of generalized industrial growth built upon the reinforcement of the industrial districts has turned into a continuous cycle of crises and restructuring compelled by global competitive pressure. The local industrial system reacts by shifting towards new segments of global markets, a process which requires new technologies, new specialities, new markets and new local leaders and hierarchies. This cycle may eventually give rise to de-industrialization, decline, or further success, but the success may lead to considerable labour saving. In any event, the outcome is highly unpredictable on both the supply and demand sides of the labour market.

The cases of Pesaro and Macerata are exemplary. Both industrial districts reached the mature stage during the 1980s, but their dynamics of employment and productivity were almost the opposite. As shown in Table 9.4, employment and productivity dynamics in agriculture and services were quite homogenous in the two cases, while they were the reverse in the industrial sector. Pesaro exhibits a large increase in industrial employment but low productivity growth; Macerata displays the opposite pattern. High industrial productivity growth as a proxy for industrial success is no longer clearly linked to employment growth. Rather, it depends on the phase of the cycle that the local system is undergoing; it depends on the specific sectors prevalent in each regional context and therefore on the specific technology and markets. These two different and divergent patterns can be partially due to the specific sectors in which the two regions are specialized. Therefore they are induced by the specific business cycles and can be reabsorbed and even inverted in the medium run. But they can also be the effect of divergent strategies in the two local economies; in the next section we will elaborate further on these issues.

Table 9.4 Income and value added (VA), 1985-1995 (millions of 1994 Italian Lira)

	Pesaro			Macerata		
	1985	1991	1995	1985	1991	1995
GDP per capita	23	25	25	25	25	27
Index (Italy =100)	97	99	96	105	100	102
GDP per worker	55	61	64	53	58	69
VA per worker						
Agriculture	21	41	38	24	31	33
Industry	58	58	61	47	59	73
Services	61	67	70	68	64	73

Source: ISTAT and Istituto Tagliacarne.

Table 9.5 Number of agrotourism farms (1998) and organic farms (1997)

	Plains	Hills	Mountains	Total
Agrotourism farms				
Pesaro	24	63	40	127
Macerata	24	63	14	101
Organic farms				
Pesaro				647
Macerata				142

Source: Regione Marche and ASSAM.

However, these processes mainly involve the core areas of both provinces, while they are less relevant in inner parts where industrial restructuring is less crucial. Here, employment in agriculture can still reach 20% and the role of public community services is crucial in providing employment opportunities. However, promising diversification strategies are increasingly taking place in both agriculture and services. Although still marginal in absolute terms, organic farming and agrotourism are two main phenomena of this diversification process.

Table 9.5 shows the number of farms devoted to agrotourism and organic farming. Considering that total farms in Pesaro and in Macerata are about 17,000 and 18,000 respectively, these are still marginal phenomena. However, they tend to concentrate in the inner parts of the provinces and are becoming the main strategic alternative for larger farms to traditional crops. Generally speaking, Pesaro seems to have a greater tendency to develop this alternative farming; this is particularly clear in the case of organic farming that is strongly biased towards the inner areas where land fertility and productivity is lower. Moreover, it has to be stressed that Objective 5b programmes have played a main role in developing both agrotourism and organic farming. This can indeed be considered the most relevant impact of these programmes in inner parts of both provinces.

Actors

As shown, the main reason of the failure of the Macerata province with respect to Pesaro in terms of employment creation within the period 1982-1995 entirely lies in the different employment effects of the industrial transformation. In fact, both experienced an intense industrial transformation and reorganization during the last 15 years induced by two main causes (Fig. 9.4):

- *Increasing international competition:* in many local strategic industrial sectors in the 1980s, a strong international competition significantly reduced the market space for the local industrial products. Typical cases are footwear and music instruments. In the former case, competition on low quality footwear by new industrialized countries forced many local firms to close or

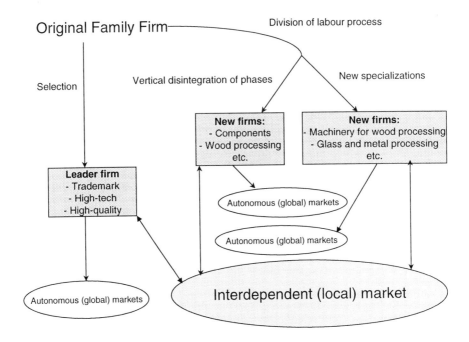

Figure 9.4   Evolutionary pattern of a complex industrial system in the two provinces

to reorient production and strategies. In the case of music instruments the competition from new electronic instruments made by Japanese producers almost destroyed the traditional local specialization in this sector.

- *Strong rationalization of the surviving firms:* to recover competitiveness on national and international markets many firms were forced to define new organizational schemes and a new market position. The main strategies are the following:

    (1) orientation to higher standard products: more quality and less quantity;
    (2) technological innovation: less labour intensive production processes;
    (3) vertical disintegration of phases: many new specialized firms coming from the original firm.

The second strategy inevitably has a negative impact on employment; the third one increases the number of firms, the flexibility and competitiveness of the local systems but has a limited impact on employment: it is a process of division of labour rather than of creation of new job opportunities. However, in terms of creation of new manufacturing firms, the balance is negative in Macerata while it is positive in Pesaro. Both show great dynamism and the difference is not determined by the number of closed firms, but rather by the number of new firms;

Table 9.6 Demography of manufacturing firms, 1988-1992

	New firms	Closed firms	Balance
Pesaro	1,525	1,356	169
Macerata	1,216	1,261	-45
Marche total	5,469	5,443	26

Source: INPS.

Pesaro shows a greater capacity in finding new markets, new niches and specialization than Macerata (Table 9.6).

To which extent this greater capacity of local actors in Pesaro is the key explanation of the different employment dynamics is not so clear. Indeed, new firms do not necessarily imply new jobs if they are created from former bigger ones. Actually, this is a quite frequent process of regenerating industrial activities during sectoral crisis and it is really the strength of these local industrial systems. However, we can say that this regenerating capacity is much better developed in Pesaro. On average, firms are smaller in Pesaro and individualism and willingness to take risks is more stressed than in Macerata; therefore, local actors show a greater capacity to react promptly to changing external conditions. Moreover, sector specialization also makes a difference. Wood-furniture production is much more complex in terms of technologically integrated production phases than footwear. On the one hand, this makes local actors able to specialize in many different 'segments' of the whole process thus creating new firms and eventually new specializations. On the other hand, this complexity makes the local system much more sound and resistant with respect to global competition. The internal complexity and vertical integration makes the competitiveness a property of the whole local system, which can very hardly be recreated elsewhere.

Characteristics of supply-side labour market

Employment growth does not depend only on the evolution on the demand side. There are also relevant processes involving the supply side of the labour market in both provinces:

- *Slow population growth:* total regional population increased very slowly in the last decade. This growth was entirely the result of increasing immigration rates (Table 9.8).
- *Low participation rates:* due to the traditional social structure the participation rate of female labour force is especially low in the inner areas. Furthermore, there is an increasing share of students within the young labour force.

Both effects implied a reduction of the working population. Moreover, in both provinces there are increasing signals of a long-term inconsistency between the features determined by the mature industrial system and by the new set of institutions, social structures and expectations generated by the process itself. The main consequences of this inconsistency became evident by the increasing mismatch between the characteristics of the local labour market and the development pattern of the local economy. In particular, the set of expectations of the young labour force can be viewed as 'urban' ones, mainly oriented either towards jobs in services (i.e. subordinate jobs with high and stable wages) or towards high-tech industrial activities. These expectations also affect educational choices, which are mainly directed towards specific university degrees, which frequently do not match labour market demand. The expectations mismatch between labour supply and demand, as well as limited access to information, eventually increase the frictional unemployment rate, particularly among the young and highly educated labour force. On the other hand, demand for low-wage and low-education labour force 'pulls' immigration from southern Italian regions and African countries (Tables 9.7 and 9.8).

The outcome of the increasing frictions between local labour demand and supply is the onset of a dual labour market (Fig. 9.5), which explains the coexistence of low global unemployment rates and high immigration rates with higher unemployment rates and emigration rates for the young highly educated labour force. Therefore, whenever the local industrial structure changes much more slowly than do local expectations and the social structure, it generates

Table 9.7 Selective unemployment

	Pesaro	Macerata
Unemployment rate, 1998 (%)		
Total	6.6	6.5
Young female labour force (age15-29)	22.2	15.3
Unemployment duration, 1992 (months)		
Total	22	17
Young graduated labour force	42	22

Source: ISTAT.

Table 9.8 Population development and selective migration, 1985/1986 and 1996/1997 (%)

	Pesaro		Macerata	
	1985-1986	1996-1997	1985-1986	1996-1997
Natural rate	-1.4	-2.3	-1.3	-3.0
Migration rate	1.2	5.1	2.3	6.0
Net rate	-0.2	2.8	1.0	3.0

Source: ISTAT.

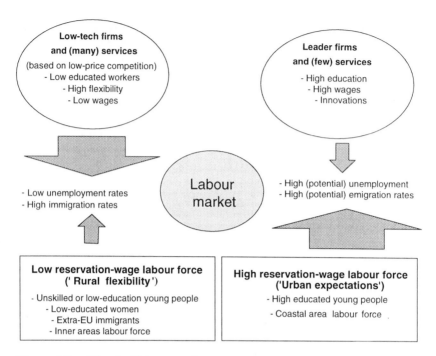

Figure 9.5   The dual labour market

pressure for a selective balance between migration and employment. Actually, the phenomenon seems to be more critical in Pesaro than in Macerata. Therefore, although the industrial transformation has induced positive effects on the employment dynamics in the leading region, its character still raises relevant issues in terms of long-run sustainability and of social change.

Key factors of success and decline in employment dynamics

The crucial process to explain success and decline in the two Italian case study regions is related to the main characters of the hierarchical reorganization of their industrial districts during 1980-1995:

- selection of best firms: the pressure on firms by global competition selected the more efficient, innovative firms;
- vertical coordination: new firms specialized in specific production phases were created to achieve more efficiency and flexibility;
- new specialization: technological progress and vertical coordination generated firms specialized in production of capital goods or high-tech components for existing firms.

This process had a positive result in the case of the furniture district in Pesaro but not in the case of the footwear district in Macerata. Besides sectoral specificities, some policies and strategies were critical in both provinces, although their mix and the specific conditions finally determined the opposite sign of the final result in terms of employment growth. These were:

- the reorganization of the industrial districts to avoid risks of excessive specialization and the creation of diversified job opportunities for a changing labour supply;
- the support and incentives to shift local industrial districts toward a higher quality standard of products and a higher technological level of processes; this shift also means the creation of a new local specialization;
- the creation of new development opportunities in rural and inner areas, which are not involved in industrial processes, but focusing on new kinds of community services, on new kinds of farming and on new specialities (handicraft and so on);
- the change in the role of the public service sector, by reducing its role and its employment capacity, but leaving increasing opportunity for new private service activities (especially linked to changing family structures that require more services both for older people, for children, and so on).

These specific conditions have played a more important role in Pesaro than in Macerata.

Besides, relevant threats seem to emerge in both regions. First, the great reduction of employment in agriculture is induced by the large pull coming from the industrial areas and is one of the causes of the increasing marginality of inner areas not involved in industrial success and of their demographic decline. This reduction has been determined by a clear tendency of farmers to reduce investments and the use of labour in agriculture to reorient resources toward other activities. The rent-seeking strategy is predominant within the region and it demonstrates the tendency to abandon agricultural activity as a source of employment opportunities and to consider it only as a source of rent. The number of professional farms is continuously reducing and their capacity to provide employment opportunities is really scarce. Second, the increasing mismatch between supply and demand of labour seems to be one of the main threats for future employment opportunities within the regions. The original family structure with its flexibility and heterogeneity progressively disappeared; young people are no longer willing to accept low wages and unsatisfying jobs. Therefore, new employment opportunities created by industries have been met only by immigrants or by people coming from inner areas, since people from coastal areas, and in particular women and young people, usually do not accept them.

Concluding remarks

The conclusions derived from the two case studies and from their comparison can be considered as lessons for policy making towards those rural regions, which show intense industrial development, low population growth and a marginal (or at least not a crucial) role of the agricultural sector in terms of employment. Four main lessons can be identified:

- Industrial success does not mean employment growth in the short-medium run; in a context of increasing global competition regions based upon small-medium highly specialized enterprises are forced into a continuous process of reorganization and technological innovation, that can imply negative effects on employment. What is really strategically important in the long run is the capacity to move towards new markets, new niches, new technologies and specialization.
- Recurrent industrial crises due to new world competitors, new technologies and changing tastes, underline the crucial role of focusing on traditional sectors and local markets. Although these are not the most dynamic ones, they can be effective in maintaining employment and avoiding dramatic impacts on regional employment. Community services, trade, agriculture and so on can play a big role when the bad times come. Therefore policies should pay attention to these sectors in good times as well.
- Employment growth and unemployment rates can be misleading when we use them as an indicator for the good performance of a regional economy and the labour market. We have to look inside the figures to see if they hide some relevant problems. In both our regions, and especially in Pesaro, young people and women show higher unemployment rates and worse employment dynamics due to increasing problems in matching local labour demand and supply.
- Our definition of rural regions can be too loose; the rurality of Pesaro and Macerata is virtual if we look at the core of both provinces. Most of the population, industrial activities and services concentrate in the core and here no traditional rural characteristics can be found. On the contrary, both case study regions are extremely rural in the inner areas where most of rural policies are actually concentrated. However, these two parts are strictly linked in terms of historical evolution and their evolutionary patterns are functional to each other. Therefore there should be a coordination, rather then a separation between the policies for the urban and the inner areas.

References

Becattini, G. (1998) *Distretti industriali e Made in Italy. Le basi socio-culturali del nostro sviluppo economico (Industrial districts and Made in Italy. The socio-cultural basis of our economic development)*. Bollati Borlinghieri, Torino.

Bernat, G.A. (1997) Manufacturing and the Midwest rural economy; Recent trends and implications for the future. *Rural Development Perspectives* 12-2, 2-12.

Esposti, R. and Sotte, F. (1998a) *Case study Pesaro*. University of Ancona, Department of Economics, Ancona.

Esposti, R. and Sotte, F. (1998b) *Case study Macerata*. University of Ancona, Department of Economics, Ancona.

Esposti, R. and Sotte, F. (1998c) *Aree rurali, società rurali e mercati del lavoro. I casi di Pesaro e Macerata (Rural areas, rural societies and labour markets. The cases of Pesaro and Macerata).* Associazione Alesandro Bartola, no. 4, Collana Appunti.

Esposti, R. and Sotte, F. (1999) Territorial heterogeneity and institutional structures in shaping rural development policies in Europe. *Proceedings of the IXth EAAE Congress on European agriculture facing the 21st century in a global context.* 24-28 August, Warsaw.

Fuà, G. (1988) Small-scale industry in rural areas; The Italian experience. In: Arrow, K. (ed.) *Proceedings of the eighth world congress of the International Economic Association The balance between industry and agriculture in economic development.* Macmillan, London.

Henry, M. and Drabenstott, M. (1996) A new micro-view of the U.S. rural economy. *Economic Review* 2, 53-70.

OECD (1994) *Creating rural indicators for shaping territorial policy.* Paris.

OECD (1996) *Territorial indicators of employment; Focusing on rural development.* Paris.

Saraceno, E. (1995) The changing competitive advantage of rural space. In: Gilg, A.W. (ed.) *Progress in rural policy and planning,* John Wiley, Chichester.

Comparison of the Dutch Case Study Regions: Drenthe and Groningen

10

Ida J. Terluin, Jaap H. Post and Anita J. Wisselink

Introduction

Selection of case study regions

The Netherlands is a small and densely populated country. Out of the 12 provinces, only five are classified as intermediate rural regions; the other regions are most urban (see Chapter 3). It is rather difficult to make an analysis of employment development in the provinces of the Netherlands since 1980 till now, due to lack of a consistent time series for this period and due to the creation of a new province (Flevoland) in 1986. Several sources for employment data for short intervals since 1980 exist, but often these time series cannot be linked to each other and sometimes show opposite trends. For the selection of case study regions we have used a Eurostat time series at provincial level on employment at the place of work, which is only available for the period 1980-1991. According to these data, within the group of intermediate rural regions in the Netherlands, the province of Flevoland had the highest non-agricultural growth rate and the province of Groningen the lowest. Flevoland consists of the so-called 'polders', areas that have been reclaimed from the Zuiderzee during the course of this century. There is a high rate of commuting between Flevoland and the neighbouring city of Amsterdam. Due to these specific conditions, Flevoland does not serve as a good example for other Dutch rural regions. Hence, we decided to select the province of Drenthe, which has the one but highest non-agricultural growth rate in this period, as the leading case study region and the province of Groningen as lagging case study region. Non-agricultural employment growth rate (measured at the place of work) in Drenthe and Groningen amounted to 4.2% and 2.6% respectively per annum compared with a national average of 2.9% per annum during the years 1980-1991 (Eurostat data).

Drenthe and Groningen not an extreme pair of leading and lagging regions

After the selection and in order to extend the period 1980-1991 to more recent years, another source of employment data was used: the provincial employment data base PWR, which is available for Drenthe and Groningen for the period 1981-1996. The problem with the time series of Eurostat and PWR is that these are difficult to compare and sometimes even show opposite results. A closer look into the PWR data shows that on the whole Drenthe and Groningen experienced more or less the same employment growth in the years 1981-1996, but that the pattern of employment growth in Groningen showed larger fluctuations than that in Drenthe (Fig. 10.1). During the first half of the 1980s Groningen had a sharper decrease in employment (5% per annum in 1981-1984) relative to Drenthe (2.5% per annum in 1981-1984). In the second part of the 1980s Groningen showed a higher annual increase in employment (over 3% per annum in 1984-1990) compared with Drenthe (2% per annum in 1984-1990). However, in the first half of the 1990s employment growth in Groningen stagnated (+0.5% per annum in 1990-1995), whereas that in Drenthe showed a moderate increase (1% per annum in 1990-1995). In 1996 employment in Groningen jumped again to a high extent, and in that year the increase in employment since 1981 was more or less similar to that in Drenthe.

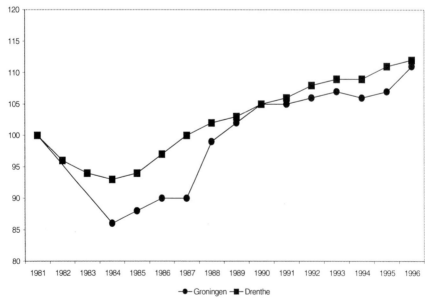

Figure 10.1   Development of total employment (at the place of work) in Drenthe and Groningen, 1981-1996 (1981=100) [a]

Sources: PWR Drenthe and PWR Groningen; adaptation LEI-DLO.

a) Data for Groningen in 1982 and 1983 are not available.

Location and brief history

The differences between the Eurostat and PWR data sources show that caution is required in the labelling of regions as 'leading' and 'lagging'. Based on the PWR figures it can be concluded that Drenthe and Groningen are not extreme cases of leading and lagging regions. Moreover, when only looking to the period 1984-1996, it appears that total employment growth in Groningen exceeds that in Drenthe. This shows that 'leading' and 'lagging' is not a permanent situation, but depends on the period considered.

Drenthe and Groningen are bordering provinces in the northern part of the Netherlands (Fig. 10.2). Together with the province of Friesland they form the so-called 'three northern provinces'. These three regions are characterized by a relatively low density of population and economic activities and are located at the periphery of the Netherlands. This common starting point sometimes results in a natural cooperation to face problems in certain fields. Drenthe has had a long history of poverty. The poor agricultural soil with low yields was often insufficient for a good living. The people often lived in turf huts and

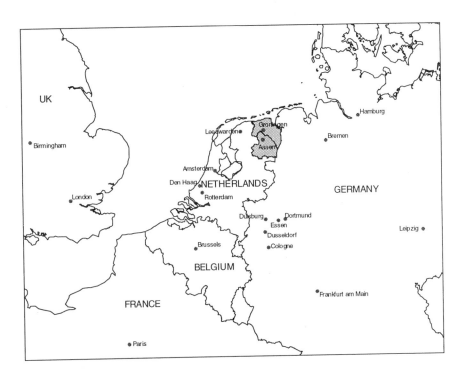

Figure 10.2   Drenthe and Groningen in a broader EU context

supplemented their incomes through peat cutting. The process of industrialization started only after World War II. Groningen has a totally different history. The region of Groningen is often referred to as 'Stad en Ommeland' (City and Surrounding), which emphasizes the regional function of the city of Groningen. Since the Middle Ages, the city has been the central market place in the region. Up to 1900 considerable areas of land were reclaimed from the sea in the northern and eastern part of the region. This resulted in colonization by non-native labourers in particularly the eastern part, who had a different mentality relative to the native population. During industrialization in the second half of the nineteenth century, various plants were built for textiles, bicycles, sugar, tobacco, coffee, tea and book printing in the city of Groningen (Kooij, 1987). At this time its service function was further enhanced, especially in the grain trade. Outside the city, activities in agro-industry (in particular starch potatoes and cardboard) and shipbuilding became important, mainly in the southeastern part of the region.

Socio-economic indicators

After World War II, over 40% of employment in Drenthe and about 30% in Groningen was in agriculture. Due to the exodus of agricultural labourers as a consequence of mechanization, shortages of employment opportunities arose. From the 1950s, the Dutch government implemented an industrialization policy for the whole country, in which it aimed at the creation of jobs in the peripheral parts of the country and a relief of the congestion of economic activities in the western part of the country. In Drenthe this policy implied that industrial development was encouraged through the granting of investment premiums, setting up industrial sites and adopting measures related to the construction of infrastructure. In Emmen and other main population centres, like Hoogeveen, Assen, Meppel and Coevorden, new enterprises were settled and the enterprises that already existed expanded substantially. In Groningen this policy induced the settlement of several big firms in the region. Moreover, the discovery of large amounts of gas in the soil of Groningen resulted in employment creation. In the 1970s the construction of the Eemshaven (port at the Eems in the northeastern part of Groningen) started. At the beginning of the 1980s, it can be said that the transformation from an agrarian economy to an industrial economy had been completed in both regions, and that a phase of decline started in the share of industries in employment with a further increase in the share of services.

From a comparison of some socio-economic indicators since the beginning of the 1980s, it appears that main differences among Drenthe and Groningen exist in the field of population growth, income per capita, the share of employment in agriculture and services and unemployment rates (Table 10.1). Due to the inclusion of the gas revenues in the per capita income, Groningen is among the wealthiest regions of the EU. However, when per capita income is corrected for this, per capita income is about 10% below the national average, but still above the per capita income in Drenthe. The unemployment rate in Groningen is much higher than in the Netherlands as a whole, which reflects a situation of a shortage of labour opportunities.

Table 10.1   Economic environment of Drenthe and Groningen, 1980-1996

	Drenthe	Groningen	The Netherlands
Population in 1996 (million)	0.46	0.56	15.49
Size of the region (km^2)	2,680	2,967	41,029
Population density (inh./ km^2), 1996	171	188	378
Population growth 1980-1996 (% per annum)	0.5	0.04	0.6
Employment growth 1980-1991 (% in persons per annum)	3.6	2.4	2.8
Population 15-65 years (as % of total population)			
1980	64	65	66
1996	67	69	68
Income per capita in ECU			
1980	8,200	17,000[a]	8,800
1993	14,500	22,300[a]	17,300
Sectoral employment as % of total employment, 1995			
Agriculture	7.3	2.3	3.7
Industries	26.5	24.3	22.6
Services	62.8	70.4	70.6
Unemployment rate (%)			
1987	9.2	14	9.9
1995	8.8	9.7	7.3

Source: Eurostat Regio Database; adaption LEI-DLO.
a) Including gas revenues.

Different distribution of cities in both regions

The composition of cities varies between Drenthe and Groningen (Table 10.2). The size of cities in Drenthe is moderate, while Groningen has some small cities but also the larger city of Groningen. In Drenthe cities are well spread over the region, whereas in Groningen the larger cities are located in a zone running from southwest to northeast (Fig. 10.3). This zoning results in a biased distribution of economic activities in the province. Due to its location near the border of Drenthe and not far from the Frisian border, the city of Groningen can be said to be the main economic centre of the northern Netherlands. In fact, many people who live in the northern part of Drenthe work in the city of Groningen. The university in the city of Groningen has contributed to the creation of a knowledge infrastructure around the city, which attracts many economic activities in the biotechnics, medical and IT sectors.

Table 10.2   Population in the five main cities, 1996

Drenthe		Groningen	
Emmen	94,000	Groningen	170,000
Assen	53,000	Hoogezand-Sappemeer	33,000
Hoogeveen	47,000	Stadskanaal	33,000
Meppel	25,000	Delfzijl	31,000
Roden	19,000	Veendam	29,000

Source: RuG, 1997.

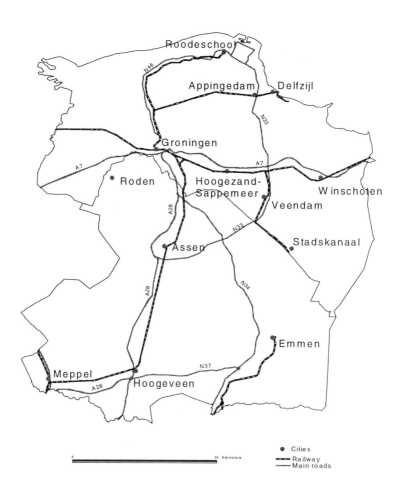

Figure 10.3   Main roads and railways in Drenthe and Groningen

Objective 2 and 5b areas

Since 1989 the eastern parts of the provinces of Drenthe and Groningen have been eligible for Objective 2 support from the EU Structural Funds. This area covers about 33% of Drenthe and 45% of Groningen and was selected as an Objective 2 region because its industry is in decline. In 1994 12 municipalities in southwest Drenthe and eight municipalities in West Groningen have been added as an Objective 5b area. This area covers about 34% of Drenthe and 25% of Groningen. According to Dutch views, both Objective 5b areas have a strong rural character with a relatively large share (about 12%) of population employed in the agricultural sector. During the last 5 years the population has declined and aged.

Local resources

Transport infrastructure rather well developed

The most important highway in Drenthe and Groningen is the A28. This highway connects Zwolle (capital city of the province Overijssel) and the city of Groningen. From Zwolle it is easy to go to the rest of the Netherlands. The province of Drenthe has started to widen the provincial road in the southeast (the N37). This route connects the A28 with the Emslandline, which gives access to the Ruhr. Another important highway in the province of Groningen is the A7, which crosses the province from west to east, going from the province of North-Holland over the 'Afsluitdijk' and the province of Friesland via the province of Groningen to Germany. The N33 is an important connection between Assen (capital city of Drenthe), Veendam and Delfzijl/Eemshaven. Several railroads (both national and international) exist, of which the railway connection Groningen-Assen-Zwolle is the most important. This connects both regions with the rest of the Netherlands. In 1995 the Rail Service Centrum Groningen (RSCG) in Veendam was opened, the largest inland container terminal of the Netherlands railways system. The RSCG is located on the route Rotterdam-Delfzijl/Eemshaven and is an important link in the intercontinental connection to North Germany, Scandinavia and the Baltic countries. Groningen has some main waterways and two seaports: Delfzijl and Eemshaven. In Drenthe there are three important lakes in the field of tourism: Zuidlaardermeer, Paterswoldsemeer and Leekstermeer. Drenthe and Groningen have a regional airport, located 8 kilometres out of the border of the city of Groningen, in Eelde (the province of Drenthe). The airport is mainly used for holiday destinations in Southern Europe and for scheduled flights between Groningen Airport Eelde and Amsterdam Airport Schiphol. However, Schiphol recently terminated these domestic flights.

Transport infrastructure in both regions is therefore well developed, which enables an efficient trade in goods and services. These roads are also an attractive location of settlement for firms, although some bottlenecks exist on the provincial road N37 in southeast Drenthe and the ringroad around the city of Groningen is liable to congestion.

Rural amenities and soil resources

Drenthe has a beautiful landscape with forests, peat bogs, moors, brooks, esdorp landscape (villages surrounded by open field) and prehistoric giant graves. These landscape amenities create an atmosphere of quietness and green space, which attracts many tourists. Drenthe is traditionally a main 'holiday province' in the Netherlands, as a large number of high quality cycle tracks give easily access to the landscape amenities. Apart from tourists, labourers, entrepreneurs and retirees are also attracted by the beautiful landscape. Accommodation and services required by these actors are a source of employment.

In contrast to Drenthe, the province of Groningen has a flat landscape with open views over extensive fields, which hardly attracts any tourists. However, its soil is a rich resource of gas, salt and magnesium chloride. These resources are processed in capital-intensive business units. Nevertheless they generate many employment opportunities, particularly the processing of salt which attracts many firms who use the salt as raw material. The soil also contains mineral water, which is used for taking cures.

Employment dynamics during 1984-1996

Drenthe has a somewhat larger share in agricultural and industrial employment than Groningen, whereas Groningen has a higher share of employment in the service sector (Table 10.1). Within the regions there is a tendency to a clustering of economic activities: Emmen and Delfzijl are centres of industrial activity, in the Meppel-Hoogeveen zone transport and distribution activities are predominant and the Groningen-Assen-Veendam zone can be labelled as a cluster of services and high-tech activities.

A detailed breakdown of employment development in nine branches can only be made for the period 1984-1996 (Table 10.3). Employment in Groningen reached a deep low in 1984 (Fig. 10.1), followed by a rapid recovery, resulting in a higher employment growth (29%) in the years 1984-1996 than that of the leading region (21%). In both regions the community, social and personal services sector generated the largest part of employment growth. The trade, restaurant and hotel sector is the second generator of jobs in Drenthe; in Groningen this place is taken by the financial services sector. Employment in manufacturing contributed in both regions to about 10% of employment growth. The financial services sector and the construction sector generated each about 10-15% of employment growth in Drenthe and in Groningen the trade, restaurant and hotel sector and the transport sector created each about 10% of the new jobs.

The relatively large employment growth in the trade, restaurant and hotel sector (including the tourist sector) in Drenthe can partly be explained by its attractive landscape, which is suitable for tourist activities. In Groningen the financial services sector grew faster than in Drenthe. An explanation for this growth is probably the city of Groningen. The city is mainly responsible for many employment opportunities in the finance, insurance, real estate and business

Table 10.3 Employment growth in Drenthe and Groningen in the different branches, 1984-1996

	Drenthe			Groningen		
	absolute growth[a]	% growth	% of total growth	absolute growth[b]	% growth	% of total growth
Agriculture	-2,186	-16	-9	172	2	0
Mining and quarrying	65	3	0	44	3	0
Manufacturing	2,861	11	12	4,221	14	10
Electricity, gas and water	103	14	0	-1,017	-42	-2
Construction	3,155	30	13	2,071	18	5
Trade, restaurants and hotels	6,097	31	25	5,136	22	12
Transport and communication	1,498	37	6	3,586	37	8
Finance and business services	3,421	44	14	13,586	132	32
Community and social services	9,437	30	39	14,696	32	35
Total employment growth	24,451	21	100	42,495	29	100
Total employment 1996	140,800			187,100		

Source: RuG, several years; adaptation LEI-DLO.
a) 1984: people who work more than 15 hours a week, 1996: people who work more than 12 hours a week; b) 1984: people who work more than 20 hours a week, 1996: people who work more than 15 hours a week.

services sector. Besides, the national government promoted employment in this sector by the relocation of government services e.g. postal and phone. A substantial part of employment growth consists of part-time jobs. This is a common feature in the Netherlands. In both regions about one-third of the expansion of employment occurred in newly established enterprises, indicating that existing companies amounted to the largest share of the total increase in employment.

Actors

Strategies of policy makers

Local and provincial policy makers tried to stimulate economic (and employment) development of their region by making the region attractive for entrepreneurs in various ways. National policy makers applied a strategy of relieving the pressure in the congested western part of the country and to persuade firms into settlement in the other parts of the country, including Drenthe and Groningen. Besides, at national level a policy of relocation of government services to peripheral parts of the country has been applied. The most important result of this policy was the relocation of a part of the national phone and post company to the city of Groningen in the second half of the 1980s, in which over 2,000 jobs were involved. The main strategies of local and provincial policy makers are discussed below.

(1) Concentration of economic activities

For a long time Drenthe has applied a strategy towards a concentration of economic activities in well-defined zones. So economic activities were encouraged in the zones Meppel/Hoogeveen/Coevorden/Emmen and Assen/Groningen. Outside the zones economic activities were discouraged. Advantages are a relatively high density of enterprises, which stimulates the interaction among enterprises and which attracts other enterprises. This leaves space for other functions (i.e. housing, recreation, agriculture and nature) in other parts of the province. However, policy makers at the municipality level, who were eager to attract some economic activities towards their municipality, sometimes thwarted this concentration strategy at provincial level.

In the province of Groningen such a concentration strategy was only officially started in 1994 with the creation of the so-called 'Stichting Groninger Bedrijfslokaties' (Groningen Business Locations) in 1994[1]. This is a cooperation of municipalities, the province, the Chamber of Commerce and the NOM (i.e. the investment and Development Company for the Northern Netherlands). The Groninger Bedrijfslokaties is now the one address in the province, which supervises the settlement of new firms in Groningen and guides them for settlement to a limited number of economic zones around Groningen/Assen/Veendam and Delfzijl/Eemshaven, which satisfy certain spatial and environmental criteria. Before 1994 enterprises were freer to select a place of settlement and there was more competition for new firms among municipalities.

In the case of the province of Groningen it can be wondered whether a concentration strategy is necessary. Outside the now defined zones the size of economic activities is limited (the main exception is Stadskanaal), and it might be expected that firms are attracted to these economic zones by infrastructural endowments and market forces. In Drenthe, where economic centres are scattered all over the region, the concentration strategy makes more sense.

(2) Strengthening cooperation between policy makers of the northern provinces

Traditionally policy makers in the three northern provinces of Drenthe, Groningen and Friesland have a close cooperation, since these provinces have a peripheral location in the Netherlands and according to Dutch standards a relatively low population density. In recent years this cooperation has been

---

[1] A similar cooperation was also created in Drenthe, the so-called Stichting Drentse Bedrijfslocaties.

strengthened and is now embodied in the SNN (Samenwerkingsverband Noord-Nederland), the cooperation of the northern provinces. The strengthening of the cooperation is the result of a growing awareness that each of the provinces is too small for its own regional development policy, and that the provinces should not compete with each other in attracting new enterprises. So the strategy is now to present the three northern provinces as one economic region, with a mix of economic activities concentrated in certain well defined zones. The advantage of such a larger economic region is that it has more critical mass than each of the separate provinces. Targets for economic zones in Drenthe are a cluster of services and high-tech activities in the zone Groningen/Assen/Veendam, a cluster of transport and distribution activities in Meppel/Hoogeveen/Coevorden and clusters of industrial activities in Emmen and in Delfzijl/Eemshaven (SNN, 1998). This zoning reflects the already existing pattern of economic centres in Drenthe and Groningen.

(3) Favourable settlement conditions for firms

In both provinces a strategy towards favourable settlement conditions for firms is applied. Land prices are relatively low in both regions. Credit facilities for SME and credit facilities for technological development are granted in all parts of the regions, whereas in some parts firms can also apply for investment subsidies (investeringspremieregeling IPR). The NOM, the investment and development company for the Northern Netherlands, aims at stimulating and reinforcing the economies of the three northern provinces in order to encourage employment opportunities. The NOM assists companies from inside and outside the region in investment matters, provides advice regarding subsidies and management, participates in financing and functions as an objective intermediary between government and enterprises. In recent years the emphasis in the strategy of the NOM has shifted from attracting new firms into supporting existing firms and to promote the settlement of supplying and processing enterprises around some big industrial enterprises. By doing so, the NOM hopes to improve the entrepreneurial climate and the anchoring of big industrial enterprises, in order to prevent these big enterprises from leaving the region.

(4) Doubling of highway N37 in Drenthe

In Drenthe, in the mid-1980s the municipalities along the provincial two-lane road N37, running from Meppel to Emmen, started a close cooperation in order to accelerate the transformation of the N37 into an international highway A37. Since part of this highway is located in Germany, German municipalities along this highway also joined the group. This cooperation of municipalities along the N37 is called the 'Stedenkring Zwolle-Emsland'. The enterprises in this area also supported the construction of the A37 and they formed the Industrial Society A37 (Bedrijvensociëteit A37). The completion of the A37 is expected within a few years and establishes a connection with the Emslandline in Germany and hence a promising international connection. The cooperation of municipalities and enterprises along the N37 is considered as successful in Drenthe.

(5) Changing attitude of municipalities and province of Groningen towards firms

Due to spatial planning and environmental criteria in the Netherlands, firms are not allowed to settle at any place in the region. These criteria were strictly applied by regional and local policy makers in Groningen in the 1980s, which resulted sometimes in firms deciding not to settle in Groningen. This strict attitude of policy makers was due to their belief in the 'makable society' (*maakbare samenleving*). A striking example of this attitude is the traffic system for the city of Groningen, as a result of which many firms became inaccessible. In the 1990s policy makers changed this attitude, for three main reasons: the society-wide tendency towards more liberalization, pressure of firms and gradual replacement of old policy makers by a new generation. Since then they have tried to solve problems in cooperation with firms in a creative way.

(6) Innovation projects in Groningen

Policy makers in Groningen implemented a strategy towards the improvement of the image of the region by launching various innovation projects. The 'Blauwe Stad' (Blue City) is a plan for regional innovation in the Oldambt in the eastern part of Groningen in order to create a high valued living and recreation environment for local people and new inhabitants. The Oldambt is characterized by a high unemployment rate, ageing population, decreasing purchasing power and a decline in the provision of services. The total area of the Blauwe Stad covers about 2,000 ha, which is at the moment mainly used for farming. According to the plans a lake of about 800 ha will be made along with the creation of about 350 ha of nature area and 500 ha forests and the construction of about 1,200-1,800 houses. It is foreseen that the Blauwe Stad will be constructed in the period 2001-2010, which requires about 4,400 person years of employment. After the construction phase, it is estimated that it will provide about 430 person years of employment (mainly in personal services, tourism, trade and transport) on an annual basis (Kleine, 1997). Total costs for the construction of the lake and the nature and forestry areas are about 125 million ECU. The plan for the Blauwe Stad is rather radical, but it is supported by policy makers of the province and the five municipalities concerned. There have been intensive contacts with the local actors in the area of the Blauwe Stad to create support, emphasizing that the plan would not be continued if the local actors would not support it. In the end local actors were positive about the plan, hoping that it will provide employment opportunities for their children.

    Other examples of innovation projects are the reconstruction of the centre of the city of Groningen and the construction of a big container terminal in Veendam (Rail Service Centre Groningen) in the 1990s. Besides, the region is promoted with the slogan 'There is nothing on top of Groningen (*Er gaat niets boven Groningen*)'. These projects show that policy makers in Groningen are innovative and not afraid of launching hazardous plans.

(7) Failure of the Eemshaven

A failure in the strategy of local and national policy makers in Groningen was the construction of the harbour area 'Eemshaven'. This was delivered during the economic stagnation in the beginning of the 1980s. No additional policies were implemented to exploit the harbour area, and it was not integrated in the regional economy. Due to these reasons this harbour failed.

Motives of firms for settlement in Drenthe and Groningen

A mix of motives can be put forward as to why firms decide to settle in the two provinces. Firstly, internal factors (pull factors) are at work, like relatively low land prices, subsidy measures, space and quietness, relatively good transport infrastructure, no congestion on the roads, the presence of buyers and suppliers of the firm and a good attitude of labourers. In Groningen firms are also attracted by the availability of raw materials (salt, gas, magnesium methyl chloride etc.), open water/ports, the availability of labourers with different educational levels, and the presence of a knowledge infrastructure around the University of Groningen and the Academic Hospital. Secondly, some external factors (push factors) affect the settlement of firms, like the congestion in the Randstad, the urbanized western part of the Netherlands. This results in a shift of companies to the middle parts of the Netherlands and also to the more northern parts like Drenthe, Friesland and Groningen, sometimes with support from the national government.

It has to be stressed that the final decision on settlement depends often on a mix of the factors above, but non-rational factors also play an important role. In Drenthe this can be illustrated by the striking example that in the 1980s a large foreign firm settled in the region because the partner of the entrepreneur lived there. New firms come from both inside and outside Drenthe and Groningen. Firms from outside come from other parts of the Netherlands (mainly Randstad and Groningen) and from abroad.

Firms leave the region for reasons of globalization of production (when the production in subsidiaries is shifted towards low-wage countries), reorganization of multinational enterprises and moving to a more central part of the country with a higher density of firms. Motives of firms for not settling in Drenthe and Groningen refer to the image and the thinness of the economic structure. The image of both regions is negatively affected as the distance between the centre of the Netherlands (Randstad in the western part of the country) and Drenthe and Groningen is considered as too large, both in physical as in psychological terms. It has to be remarked that distance is a relative notion: the distance between the Randstad and Drenthe/Groningen is about 200/250 kilometres or about 2-3 hours travelling by car. The thinness of the economic structure prevents the settlement of firms, who prefer the proximity of a large number of other firms, in order to have the opportunity of a tight network and to have supplying and processing firms in the neighbourhood and a sufficient variation in the supply of labour (which is available due to the presence of a large number of companies).

Networks are small and surveyable

Firms participate both in networks with other firms and in networks with policy makers. Internal networks are small, surveyable and characterized by easy communication. Drenthe and Groningen are small regions and local actors know each other. On the whole, the functioning of the networks is assessed to be reasonable or good. However, the weaknesses of the small networks are that actors are not very critical of each other, in fear of losing unity among actors, that actors are too inward looking and that the density among actors is too low. As a result of the low density, contacts between actors have the character of sociability instead of an incentive to innovate, which often occurs in 'more stressed' networks, i.e. networks in an environment with a high density of actors and a high frequency of contacts. With regard to this aspect, the functioning of the networks could be improved. On the whole, networks are insufficiently directed towards actors outside their own region. An improvement in this sense is the strengthening of the cooperation of policy makers in the three northern provinces in recent years.

Entrepreneurs in Groningen are active in increasing the density of actors in their business sites. AKZO Nobel advertizes in its mission that 'AKZO Nobel Delfzijl is an attractive place of settlement for business units, joint ventures and other firms'. Firms in Veendam pushed for the establishment of the Rail Service Centre Groningen and entrepreneurs in knowledge intensive activities in the city of Groningen created an attractive atmosphere for other firms. In Drenthe entrepreneurs along the N37 road try to create an attractive location for firms.

In Drenthe networks in the tourist sector are rather weak due to an insufficient willingness to cooperate. As a consequence of the increase in leisure time, the demand for tourist and recreation facilities has been increased. In order to use this opportunity, the supply and quality of accommodation has to be continuously updated. However, the level of most tourism accommodations in Drenthe is relatively low and the weak networks discourage its upgrading.

Local leaders from inside and outside the region

Within the networks some local leaders can be identified: actors who are able to push and to activate other actors to a large extent. In Drenthe it is striking that local leaders within the group of entrepreneurs often are of non-Drenth origin. This can be explained by the cautious attitude of Drenth people, which partly originates from the long tradition of inferior peat labour and small peasant farming in the region. As immigrants usually tend to do, these newcoming entrepreneurs have the capability to mobilize other local actors, probably due to the fact that their attitude differs from the local actors. This means that the functioning of the networks could be improved by the attraction of external entrepreneurs. In Groningen the situation of local leaders is different: due to its long tradition of industrial entrepreneurship, local leaders in the networks of entrepreneurs consist both of native Groningen entrepreneurs as well as entrepreneurs of non-Groningen origin.

Capacity of local actors to innovate

Due to the dependent attitude of people in Drenthe and Groningen, their capacity to innovate is rather low. However, when they are convinced by other people of an innovation, they put their doubts aside and embrace the new ideas. New coming people into Drenthe often act as innovators; in Groningen innovating local actors mainly originate from the groups of young people, policy makers and entrepreneurs.

Engine of employment growth is driven by internal or external forces

The dynamics of a region comes both from local actors and external actors. In both regions the clustering of economic activities and the abundant space can be considered as endogenous forces. In Drenthe the cooperation of municipalities and firms in the field of the widening of the N37 and the attractive living circumstances can also be referred to as an endogenous force for development. External forces for both regions are the congestion in the Randstad. As a consequence firms are pushed to other regions in the Netherlands.

Key factors of success and decline in employment dynamics

As neighbouring regions Drenthe and Groningen have much in common with regard to the attitude of actors, policies etc. Nevertheless, some striking differences (or key factors of success/decline) between both regions can be indicated, which affect employment dynamics. These are discussed below.

Key factors of success in Drenthe

(1) Attractive landscape

The beautiful varied landscape in Drenthe of forests, peat bogs, moors, brooks, esdorp landscape, agricultural landscape and giant graves creates an atmosphere of quietness and green space, which attracts tourists, labourers, entrepreneurs and retirees. The accommodation and services required by these actors are a source of employment. Moreover, the newcoming entrepreneurs also create employment opportunities in their own businesses.

(2) The neighbourhood of the city of Groningen

The city of Groningen boosts employment in the northern part of Drenthe. Groningen is the main economic centre in the north of the Netherlands, which is located a few kilometres from the Drenth border. Many companies look for settlement opportunities in the neighbourhood of this city, and often settle in the north of Drenthe. This is a main factor in the creation of employment in the northern part of Drenthe. Due to the attractive housing circumstances in this part, many labourers, who work in the city of Groningen, prefer to live in the northern part of Drenthe.

(3) Favourable business site conditions

Land prices for new establishments in Drenthe are relatively low compared with the more congested western part of the Netherlands (Randstad). Moreover, facilities like investment subsidies and favourable fiscal treatments and the relatively low prices of real estate also enhance the attractiveness of Drenthe for entrepreneurs looking for a location to settle. Drenthe has applied a concentration policy for locations of settlement. This has two advantages: within certain zones there is a clustering of economic activities, which enhances the density of entrepreneurs and the opportunities for networking, and it contributes to the safeguarding of the attractiveness of other zones for other functions.

Key factors of success and decline in Groningen

(1) Concentration of employment opportunities

In 1997 about 43% of employment was located in the city of Groningen and another 27% in the smaller towns of Hoogezand, Veendam, Winschoten, Stadskanaal and Delfzijl. This implies that most economic activities are located in a southwest-northeast corridor in the region, and that there are few employment opportunities in the northwestern and southeastern parts. This concentration has positive and negative effects:

- Positive: the concentration of economic activities enhances the density of actors in the networks.
- Negative: in certain parts of the region, and in particular in the eastern part, there are too few employment opportunities for less educated labour. On the whole the willingness to commute a long distance among less educated people is relatively low. This implies that the young less educated, more mobile people leave that part of the region and look for a job elsewhere, whereas the older less educated people stay. Among this group there is high unemployment.

(2) City of Groningen

Groningen is the main economic centre in the north of the Netherlands. However, the city is not centrally located in the region, which implies that the neighbouring provinces of Drenthe and Friesland also profit from the city. The city of Groningen is considered as a nice city with a lot of employment. This attracts people and causes spillover effects. The university is one of the largest employers in the city. This has contributed to the creation of a knowledge infrastructure around the city of Groningen, which attracts many high-tech activities. The importance of this infrastructure for employment has increased since 1980, for example due to the establishment of the Zernike Science Park and the extension of activities in biotechnics, the medical sector and the IT sector. On the other hand, the city of Groningen has a high unemployment rate, as many young people, who have completed their studies, but have not yet found a job, tend to stay in the city because of its attractiveness.

(3) Peripheral location

A main distinction between Groningen and Drenthe refers to the location of Groningen. Of the three northern provinces Groningen has the most peripheral location relative to the economic centre of the Netherlands (Randstad). So Groningen will profit less from the outflow of firms from the congested Randstad. Therefore, Groningen has to strengthen its economy (and so its employment opportunities) more by its own power. However, in order to compensate for this disadvantage, Groningen is supported by national policy makers, who have implemented a policy towards stimulating regional economic development by means of a relocation of government services to peripheral parts of the country. In this way over 2,000 jobs were involved in the relocation of a part of the national phone and post company to the city of Groningen in the second half of the 1980s.

Concluding remarks

From the analysis of employment dynamics in the neighbouring regions of Drenthe and Groningen, it appeared that they have much in common and that it is hardly possible to label Drenthe as 'leading' and Groningen as 'lagging'. We can learn the following lessons for employment creation in rural regions from these case studies:

(1)   Concentrate economic activities
The clustering of economic activities increases the density of firms and the opportunities for interaction and linkages between firms. These agglomeration effects may result in a dynamic development process. Besides, it leaves space for other functions like housing, recreation, agriculture and nature in other parts of the region.

(2)   Provide good infrastructure
The problems with the two-lane road (N37) in Drenthe are currently being solved by a successful cooperation of entrepreneurs and policymakers. This road allows for an efficient transport of goods and services and is an attractive location for settlement.

(3)   Integrate infrastructure into a broader perspective
One of the reasons why the harbour area Eemshaven failed was that it was constructed as a 'cathedral in the desert', with no links with the surrounding economy. So infrastructure projects should be included into a broader territorial development concept.

(4)   Permanent upgrading of tourist accommodation
Tourists can select innumerable destinations. In order to satisfy the demands of tourists, the quality of tourist and recreation facilities has to be continuously updated.

(5)   Cooperate with neighbouring regions
In order to create more critical mass, such a cooperation can be useful. Neighbouring regions can present themselves as one large economic area and apply a joint regional development strategy. An agreement that the participating regions work together in attracting firms to the area instead of competing with each other for firms, can be part of such a cooperation.

(6)   Cooperative attitude of policy makers towards firms
Firms can face various problems with regard to spatial and environmental issues. These should be solved together with policy makers in a creative way.

References

Kleine, J. (1997) *De Blauwe Stad; Van idee naar werkelijkheid; Eindrapport Stichting Blauwe Stad (Blue city; From idea to reality; Final report Blue City).* Groningen.

Kooij, P. (1987) *Groningen 1870-1919; Sociale verandering en economische ontwikkeling in een regionaal centrum (Groningen 1870-1919; Social change and economic development in a regional centre).* Van Gorcum, Assen.

RuG (Rijksuniversiteit Groningen/State University Groningen) (1976-1997) *Statistisch jaarboek voor het Noorden (Statistical yearbook for Northern Netherlands).* State University Groningen, Faculty of Spatial Science, various issues, Groningen.

SNN (1998) *Kompas voor de toekomst; Ruimte voor de ontwikkeling van Noord-Nederland; Reactie op het advies van de Commissie-Langman (Compass for the future; Scope for development in the Northern Netherlands; Comments on the report of the Langman Commission).* Groningen/Leeuwarden/Assen.

Terluin, I.J., Post, J.H., Wisselink, A.J. and Overbeek, M.M.M. (1999a) *Forces affecting employment dynamics in Drenthe; Case study in a leading rural region in the Netherlands.* LEI-DLO, The Hague.

Terluin, I.J., Post, J.H., Wisselink, A.J. and Overbeek, M.M.M. (1999b) *Forces affecting employment dynamics in Groningen; Case study in a lagging rural region in the Netherlands.* LEI-DLO, The Hague.

Employment in Rural Austria: the Case Study Regions Osttirol and Liezen

11

Franz Weiβ

Introduction

This chapter gives a brief summary of the two case studies in Austria: Osttirol, as an example of a leading rural region, and Liezen, as an example of a lagging one. The results are based both on the analysis of data and literature and on interviews with 60 entrepreneurs and 15 representatives of relevant institutions (held between September 1997 and March 1998).

Selection of the case study regions

The case studies were selected according to non-agricultural employment growth for the period 1981-1991. In Osttirol the rate of non-agricultural employment growth p.a. was more than 0.5% above the Austrian average for this period, while in Liezen it was almost 0.8% below. Both regions were not the most extreme cases. So the growth rates of non-agricultural employment in the two regions Mühlviertel and Tiroler Unterland were higher than in Osttirol, and the region Westliche Obersteiermark developed worse than Liezen. However, Mühlviertel is closely located to one of the five largest cities of Austria. Since the good development of peri-urban regions was a general trend in the last 15 years, and there are no peri-urban regions with a poor development in Austria, we cannot expect lagging peripheral regions to adopt their strategies successfully. Tiroler Unterland has a long industrial tradition with old locally rooted firms. In economic terms, it had already been far above the average before 1980 (the starting point of our study), and it seems hard for a lagging region to copy the development if it faces less promising preconditions. In contrast, Osttirol has been one of the most lagging regions of Austria for decades. In 1980 a period of strong industrial development started, and lasted until the beginning of the 1990s. Even if income is still below the national average and unemployment is high, the development was remarkable. Therefore, we have chosen Osttirol for the case study in a leading rural region in Austria. With regard to the selection of a lagging case study region Westliche Obersteiermark and Liezen were candidates, two neighbouring regions belonging to the same province. We have chosen Liezen for its extraordinarily high unemployment rate and its similarity to Osttirol.

Basic information on the case study regions

Osttirol is located in the southwest of Austria (bordering to Italy) and it is part of the province Tyrol (Fig. 11.1). It has about 50,000 inhabitants and an area of 2,000 square kilometres. Due to the alpine location, settlement is more or less confined to the main valleys. The most important local resource is the beauty of the landscape, which attracts tourists both in summer and winter. In addition, extensive forests and alpine pastures are an important resource for economic activities. We can divide Osttirol into three parts: the basin of Lienz in the southeast, the Drau-valley in the southwest and the north of the region (Fig. 11.2). Population and employment are extremely concentrated in the regional centre Lienz. So, 48% of population lives in the basin of Lienz, and 63% of employment is concentrated there. In contrast to most valleys in the region, which are usually very narrow and therefore not appropriate for industrial expansion, the basin of Lienz is wider, since two rivers join there. This makes the basin of Lienz an attractive location for industries and business services. The other two parts of the region both have a share of about a quarter of the population and below 20% of employment. While Matrei in the north has exactly the same share in population and employment, Sillian-Heinfels in the southwest faces a positive commuting balance, which demonstrates its function as centre of a small local labour market.

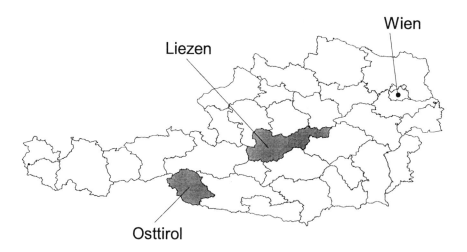

Figure 11.1   The location of Osttirol and Liezen in the Austrian context

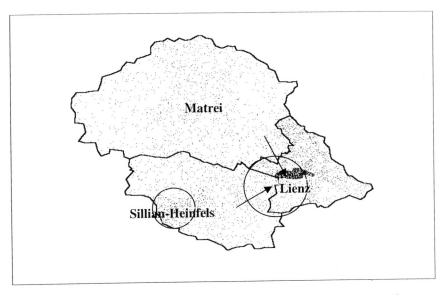

Figure 11.2   Distribution of population and community centres in Osttirol [a)]
a) Dots indicate population density and arrows flows of commuters.

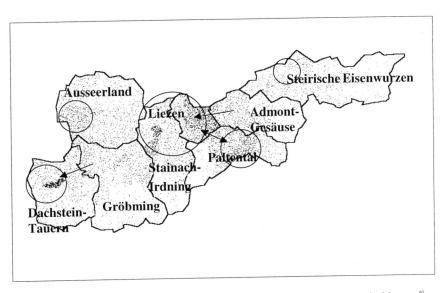

Figure 11.3   Distribution of population and main community centres in Liezen [a)]
a) Dots indicate population density and arrows flows of commuters.

For historical reasons, Osttirol, has no common border with the rest of the province, and until 1968 the only reasonable road-connection to the provincial capital Innsbruck crossed Italy. The border location, the long distance to any urban centres, and the bad road connection to the rest of the country has hampered the economic development for decades. Due to the creation of the Felbertauern road in 1968, the situation has improved and the good economic development of the 1980s can be partly explained by this fact.

In 1981 the employment share of agriculture was 16%, compared with 34% in industries and about 50% in services. Since the 1980s, employment in both industries and services increased substantially, while agricultural employment decreased. Despite a strong increase in jobs, unemployment grew massively during the 1980s and 1990s, and the unemployment rate is high compared with other rural regions (Table 11.1). However, the dominant share of unemployment is seasonal, due to tourism and construction, and therefore not necessarily involuntary.

Liezen is located in the geographical centre of Austria and belongs to the province Styria. It has 80,000 inhabitants, and an area of 3,200 square kilometres. Similar to Osttirol, there are no urban centres within a distance of 100 km, but the road connections are generally better. Since the late 1980s, there is a motorway, which connects the region to the north and the southeast of the country. A special problem is the connection to the west, since the congested road creates an interest conflict between commercial transport and the population. The endless struggle for a solution led to a temporary speed limit, frequently lamented by local firms.

Like Osttirol, Liezen is an alpine region with settlement in the valleys. However, in contrast to Osttirol neither population nor employment is significantly concentrated in a regional centre. The four largest towns of Liezen do not have a much higher share of population, and even a lower share of employment, than the regional centre of Osttirol (Lienz) alone. Liezen can roughly be divided into three main parts: the west, the centre and the east (Fig. 11.3). The west (Gröbming, Dachstein-Tauern and Ausseerland) is the tourism zone, while the centre (Liezen, the Paltental and Stainach-Irdning) is the industrial zone. Finally, the east (Admont-Gesäuse, and Steirische Eisenwurzen) is a remote part. The central and western part of the region each have a share of over 40% in population, whereas only 16% live in the east. Similarly, 47% are employed in the centre, 40% in the west, but only 13% in the east. The most important local resources are the beauty of the landscape (especially in the west and the east), mineral resources (salt, plaster of Paris, marble, talcum), extensive forests and alpine pastures, and numerous hydropower plants.

In 1981 the employment share of agriculture was 13%, compared with 39% in industries and 47% in services. This reflects the long industrial tradition of the region. Since 1980 employment development was significantly worse than in other rural regions. Agriculture and industries both faced a strong decline of employment, while services employment increased, but below average. The unemployment rate of Liezen is one of the highest in the country, and has increased by a factor six since 1980. Unemployment is marked by a high share of

Table 11.1 Basic indicators for Osttirol and Liezen

	Osttirol	Liezen	Most rural regions	Intermedia-te regions	Most urban regions	Aus-tria
Population, 1991 (1000)	48.3	81.4	3,798	2,458	1,540	7,796
Size (1000 km^2)	2	3.2	67.4	14.7	0.4	82.5
Population density, 1991 (inh./ km^2)	24	25	56	167	3,879	94
Population growth, 1981-1991 (% p.a.)	0.2	0.1	0.3	0.5	0.1	0.3
Total employment growth, 1981-1991 (% p.a.)	0.6	-0.4	0.2	0.9	0.3	0.4
Non-agricultural employment growth, 1981-1991 (% p.a.)	1.3	-0.04	0.8	1.0	0.3	0.8
Sectoral employ-ment structure,[a] 1991 (%)						
Agriculture	10	10	12	3	1	6
Industries	35	34	38	36	28	35
Services	55	56	49	61	71	59
Unemployment rate (EU method),[b] 1991 (%)	2.9	4.4	3.2	3.4	5.2	3.7
Unemployment rate (Austrian method)(%)						
1981	3	2.1	2.1	1.7	2.3	2
1995	8.5	9.4	5.4	4.9	7.6	5.7
GVA/capita, 1991 (ECU)	11,885	12,145	12,160	18,670	23,8890	16,530
GVA/capita growth, 1981-1991 (% p.a.)	8.5	5.7	7.5	7.5	8.8	7.8
Population share of the main town, 1991 (%)	25	9				

Source: ÖSTAT.
a) Agriculture: agriculture and forestry; Industries: mining, manufacturing, electricity, gas and water, construction; Services: trade, restaurants and hotels, transport, storage and communication, finance, insurance, real estate and business services, community, social and personal services; b) According to the EU method unemployment figures are based on representative enquiries; according to the Austrian method they are based on the average number of registered unemployed persons. The values can differ substantially, because especially in tourism and construction work is confined to certain seasons. For the rest of the year a person working in those branches can receive unemployment benefits and is therefore registered as unemployed. However, according to EU definitions he is not unemployed, and therefore filtered.

long-term unemployed and a high share of unemployed women.

In the maps of the case study regions, the dots indicate population densities (at the level of communities)[1], the arrows show significant flows of commuters (1991), and circles indicate regional labour markets (Figs 11.2-3). Osttirol is an Objective 5b region with 100% Less Favoured Area (LFA) (mountain area). The western part of Liezen is under Objective 5b as well, while the eastern part is under Objective 2. Liezen is also as a whole classified as LFA (mountain area), due to the extreme conditions for agricultural production.

Economic activities

Osttirol shows an above average development both in employment and productivity in the 1980s, while Liezen is below average in both (Figs 11.4-6). The sectoral diagrams make clear that the difference in employment growth of Osttirol and Liezen is due to industries. Services developed very similarly, and in agriculture employment decreases were even lower in the lagging region than in the leading one.

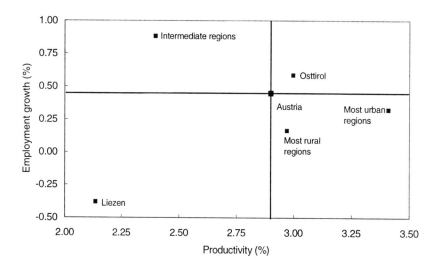

Figure 11.4   Total employment and productivity growth, 1981-1991 (% p.a.)
Source: Östat, Volkszählung 1981, 1991.

---

[1] Notice that this yields only a coarse picture of the actual distribution. In fact population is much more concentrated in the main valleys, but communities are usually large in terms of land, and so in the map the population of a town or a village is distributed over the whole area of the community.

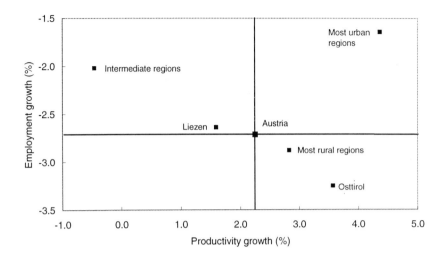

Figure 11.5   Employment and productivity growth in agriculture, 1981-1991
       (% p.a.)
Source: Östat, Volkszählung 1981, 1991.

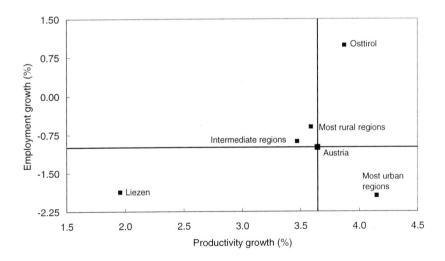

Figure 11.6   Employment and productivity growth in industries, 1981-1991
       (% p.a.)
Source: Östat, Volkszählung 1981, 1991.

In order to get a clearer picture of the development, one has to go deeper into the level of economic branches. The employment shares of the branches are very similar in both regions. The main difference in the rural average is a lower share in manufacturing, and a higher share in accommodation. However, this is not surprising, since both regions are alpine regions with a high importance of tourism. Figure 11.7 relates employment growth (1981-1991) to the average of rural regions in Austria according to economic branches. It shows that the difference between the two regions is more or less explained by the development in manufacturing and construction. In services the better performance of Osttirol in public services and accommodation was more or less balanced by a worse development in transport and trade. In agriculture the lagging region experiences a smaller decline in employment than the leading one. This can be explained by two facts. First, the strong food manufacturing industry in Liezen provides better sale conditions for farmers than is the case in Osttirol, where just few agricultural products are manufactured within the region. Second, in Osttirol incentives to leave the agricultural sector are higher than in Liezen, since employment opportunities in other sectors are better.

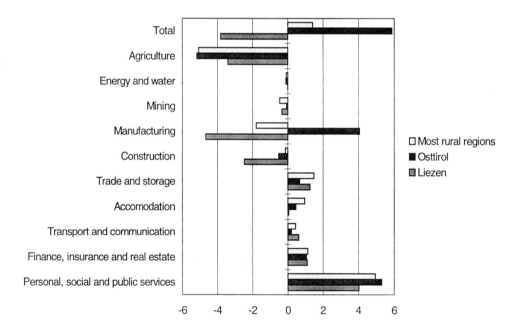

Figure 11.7 Employment growth by branche, 1981-1991 (%)
Source: Östat, Volkszählung 1981, 1991.

In Osttirol the main contribution to employment growth in the 1980s came from firms in metal manufacturing. Three firms settled between 1978 and 1980, and created about 1,500 new jobs up to 1990. The largest one is part of a big company with German ownership, and produces refrigerators. Other important contributions came from wood manufacturing and public services. While employment in saw mills was rather on the decline, the improved development of wood manufacturing was mainly caused by increases in the production of furniture, usually in small and medium-sized local firms. In public services employment increases are due to public institutions. The hospital grew from 450 to 600 jobs, and schools increased employment by 110 people.

As in the 1980s, employment development in Osttirol was above the rural average in the 1990s. In contrast to the 1980s, between 1991-96 the construction and services branches were the main creator of jobs. While in nearly all manufacturing branches jobs were lost, a large number of jobs were created in construction and almost in all services, especially in trade, medical and social services. In trade the increase consists of a large number of part-time jobs. In medical and social services the further expansion of the hospital and the home for the aged was the most relevant employment factor. The employment increase in construction was mainly due to public projects.

While in Osttirol metal manufacturing was the engine of growth during the 1980s, it was metal manufacturing, which was mainly responsible for the lack of development in Liezen. Two large firms reduced employment by 1,700 jobs, and one medium sized firm (100 employees) left the region for cost reasons. One of the two large firms had been part of the formerly state-owned metal industries VOEST, and has been guided by different private owners since 1992. The second one had been part of a big German company until 1982, then temporarily managed by a local initiative, and finally sold to German industrialists in 1988. Partly, the decreases in metal manufacturing were balanced by a new large enterprise, which settled in the 1980s, and has created 470 new jobs since then. Other substantial job decreases happened in manufacturing of non-metallic minerals and shoe production. A large manufacturer of magnesite reduced employment from 800 to 280 jobs since 1980 due to rationalization and restructuring. This was partly balanced by the employment growth of two manufacturers of plaster of Paris. The employment decrease in shoe production is due to one large firm, which left the region for cost reasons in the 1980s. It caused a loss of 270 jobs, mostly for female employees. Finally, jobs were reduced in numerous medium-sized construction firms and the salt mine in the northwest.

Job increases in Liezen can be found in services, but only in the transport branch were those increases above average. This was mainly due to the public railway company, which is strongly present in the region, but also to some large firms in road transport, which have benefited from the new motorway.

In both regions the big private investment projects came predominantly from outside. However, local actors were very active in attracting those new firms, and so it is hard to say whether the engine behind employment growth was internal or external to the region.

Actors, policies and strategies

Institutional issues

Austria is divided into four administrative levels: the federal state, nine provinces, 121 districts and more than 2,000 communities. Osttirol and Liezen are both districts. The legislative power is concentrated at the federal state and the provincial level. Administration follows the principle of subsidiarity. National and provincial parliaments are elected directly by the population, while governments are elected indirectly by the parliaments. So, political decisions are usually not drawn at district level, but rather at national or provincial level. Similarly, local affairs are usually decided in municipalities and not at district level. The local representatives in national and provincial parliaments are an important connection to the provincial and national authorities. They usually form the bridge between local politicians and provincial and national ones.

The most important local networks are the interest groups like the chambers of commerce, labour and agricultural organizations. At the local level the chambers are primarily engaged in lobbying, providing information and education programmes, and generally helping their members. Frequently they have a mediating role between politicians, entrepreneurs and labourers. The chambers are linked together in the provincial and the national chambers, and so are the affiliates of the Labour Market Services. They all have democratic structures, and important decisions (i.e. positions in tariff-negotiations) are taken in big meetings. Provincial representatives are elected directly by the members (firms, labourers, farmers). Other formal networks of local enterprises with external enterprises exist in tourism and among the affiliates of companies.

Formally, local politicians and interest groups work together in several regional committees. Some of them are permanent, while others are temporarily connected to specific projects. However, due to low population density, people generally know each other, and know what happens in the close neighbourhood, and so many things happen on the basis of personal relations.

Generally, local actors assess the functioning of networks to be similar to other regions. The cooperation between local politicians and entrepreneurs is generally good in Osttirol, while in Liezen it is good in the east, but bad in the centre, which is the industrial zone, and the west.

Policies and strategies

In both regions employment in the public sector has been expanded. In Liezen expansion was below average, while it was slightly above average in Osttirol. In both regions employment increases were concentrated in medical and social services, and in Osttirol also in education.

Public investment projects like the creation or expansion of public buildings, sewerage systems or roads, usually have an important impact on employment in construction. In the period of study, this was especially the case in Osttirol where many smaller projects have been carried out. In contrast, in Liezen local firms could hardly benefit from big projects like the creation of the motorway.

In Austria direct labour market policy is administrated by local affiliates of the Labour Market Service. This institution mediates jobs, administrates unemployment benefits and financial supports of the *Arbeitsmarktförderungs Gesetz (AMFG)*, and is engaged in education programmes for unemployed people. In both regions the strategy of job mediation has changed since 1980. While in the 1980s each unemployed was allocated to one mediator, for a few years there has been a service for entrepreneurs and a service for people looking for a job. Both services are closely linked via the computer system. This guarantees flexibility in mediation, close contacts to firms, and is more appropriate for large numbers of unemployed people.

In the field of training, in Osttirol a technical school (ISCED 3) was created some years ago, while in Liezen little has happened since 1980. In both regions the chamber of commerce created schools for technical education in electronics and machinery. In addition to that, there are education programmes for unemployed people, paid by the Labour Market Service, and education programmes run by the chambers. Usually, they are strongly oriented to the local needs. However, this was similar in both regions.

In Osttirol local politicians played an important role in attracting new firms from outside. Active communities created good preconditions for industrial development (industrial zones with necessary infrastructure), used personal contacts to get in touch with firms, and then showed cooperative behaviour, when firms were interested. So firms were persuaded with cheap land, cheap connection to energy, water and canalization, and generous supports in the starting period.

In Liezen this was only the case in the east, while local politicians in the centre and the west were described as uncooperative towards industrial firms. Partly this has to do with the fact, that the capital Liezen wants to emphasize the development of trade, instead of industries, and the west is specialized in tourism. In fact, the capital Liezen developed very well as a trade centre of the whole region. This development was initiated by a private initiative of local entrepreneurs, which systematically tried to attract new trade firms.

In both regions the emphasis on infrastructure policies was in tourism and the road network. In tourism the creation and expansion of accommodation facilities and ski lifts was supported by direct payments and cheap loans. In Liezen the most important road project was a motorway, which connects the region from north to southeast, and reduces the travel time to urban centres like Linz and Graz. A motorway to the west has been prevented by ecologists until now. In Osttirol no new roads have been created in the investigation period, but the creation of the Felbertauern road in 1968 had still strong influence on the development during the 1980s.

Financial support for projects and enterprises came predominantly from national and provincial funds. Three groups can be distinguished. First, there are economic funds, which can be utilized by all firms meeting certain requirements. They are not tied to backwardness or geographical disadvantages of the region, even if in some cases firms in disadvantaged regions can receive higher grants. Second, there are regional funds, obtainable only by firms in disadvantaged regions, and finally, some regions have local funds, which are exclusively devoted to those specific regions, and where, in contrast to economic and regional funds, local authorities take part in the decisions.

While the availability of economic funds was similar in both regions, the main difference was in regional and local funds. In Liezen local funds did not exist, and all decisions about the use of regional funds were exclusively drawn at the provincial or national level. In contrast, in Osttirol, the regional council (a council of local policy makers) played an important role in decisions about the use of provincial funds. The emphasis was on support for tourism projects (ski lifts, bicycle tracks and a national park) and support for key enterprises. Two programmes were of importance in Osttirol. First, the so-called *Raumordnungs-Schwerpunkt-Programm* of the province Tyrol, which follows unique guidelines for all Tyrolean districts, but with better conditions for Osttirol. From the regional programme Osttirol received about 1 million ECU per year on average. Second, in the 1980s there was a special support programme for Osttirol *(Sonderförderungs Programm Osttirol),* which was co-financed by provincial and national funds. Expenditure figures for this special support programme were not available. Financial support of firms (for foundation, expansion or investments) by both funds was frequently tied to some minimum size and the obligation to create new jobs. Most of the larger firms, which contributed massively to employment growth, received substantial support from those regional funds.

Key factors of employment dynamics

Osttirol

As can be seen from the previous sections, the major reason for the above average development in Osttirol was the industrial development process in the 1980s, in which the settlement of a few large metal-manufacturers and the expansion of some local small and medium sized enterprises in wood manufacturing played an important role. Several factors have significantly contributed to this industrial development:

(1) Well educated labour force

Flexibility and problem-solving capacity is one of the potential strengths that makes firms in rural regions competitive against firms in low-wage countries. This requires labourers with medium-level technical skills. In Osttirol skills of this kind are available, and the willingness for education and permanent training is high. This was emphasized by almost all managers as an extraordinarily important argument for new enterprises.

(2) Low labour unit costs

Industrial labour costs in Osttirol were only two-thirds of the national average in 1981. Even if productivity was lower than on average, this was a substantial cost advantage for new firms that settled in the region.

(3) Cooperative policies

Communities showed cooperative behaviour, when firms indicated an interest in settling in the region. Connection costs for water, energy and canalization were kept low, and land was provided cheaply. Generous financial supports for new enterprises came especially from the province, but also from national funds. Moreover, cheap loans, interest support and direct payments for other enterprises were generally higher than in other rural regions. Finally, financial grants were frequently connected to the obligation to create new jobs.

(4) Felbertauern road marks end of isolation

Since World War I, Osttirol was an isolated region, as the only connection to the north and the rest of the province Tyrol went through Italy. In 1968 the Felbertauern road was opened, which connects Osttirol to the neighbouring region Pinzgau/Pongau, and opens new exporting possibilities to the north. With some delay, this prompted the late industrialization process of Osttirol in the 1980s.

Liezen

Contrary to Osttirol, it was industries that caused the negative employment development in Liezen. The highest employment reductions occurred in metal manufacturing, followed by the manufacturing of non-metallic minerals and the production of shoes. The following factors are supposed to be crucial for this development:

(1) Large inflexible enterprises in basic industries

At the outset of the 1980s, the industry structure of Liezen was characterized by few large industrial firms. The employment shares of basic industries (metal, leather, non-metallic minerals) was very high for historical reasons. Since those branches developed very poorly in all parts of Europe, employment decreases can partly be explained by the industry structure. Moreover, the wage level[2] was significantly above the rural average, and this was not balanced by a higher productivity. Declining prices on world markets would have required strong measures of rationalization and a continuous change towards new products at an

---

[2] The Austrian system of labour contracts makes it very difficult for a firm to reduce actual wages, since the Labour Union does not only negotiate the growth rates of minimum wages, but also the growth rates of actual wages. At the beginning of the 1980s the wage level in Liezen was traditionally high in the large industrial firms due to the good economic development in the 1960s and 1970s. When prices decreased due to strong competition on the world market, and generally sales declined, it was impossible to adapt the wage level to the new situation. Moreover, the firing of people created big compensation costs, partly borne by the state. Finally, job reductions in the state-owned company were hardly possible for political reasons. This resulted in a strong inflexibility and tardy reforms, and at the end has worsened the intensity of the crises. In contrast, wages in 1980 have been low in Osttirol, since industries had hardly existed there before.

early stage (in the 1970s). However, this did not happen due to bad management, inflexible labour contracts, and political pressure (especially in the state-owned company). During the good times of the 1960s and 1970s, nobody wanted to believe that those big ships could ever sink. So, reforms were delayed, and this has dramatically worsened the situation at later stages.

(2) Low local entrepreneurial potential

One of the major problems of regions with large industrial firms can be seen in the fact that the local entrepreneurial potential will is low. This has three main reasons. Firstly, since wages in industries are high, incentives to take the risk of being self-employed are low. Secondly, it is hard for small firms to compete with the high wage level in industries. Finally, the upward and downward linkages of large industrial firms are generally low, since they are flexible, and buy where it is cheapest. This continuously reduces entrepreneurial capacities, and can destroy the viability of a region for decades. In case of a crises no young entrepreneurial characters are present.

(3) Weak policies

The main weaknesses in regional policies are threefold. Firstly, except for the east of the region, only little has been done in Liezen in order to attract new industrial firms at local level and provincial level. Most activities of local politicians have rather been devoted to subsidizing the old sinking ships. Secondly, provincial policies in Styria are much more devoted to the capital Graz and the surrounding regions, than to the distant regions in the north, like Liezen. Thirdly, there is a lack of regional consciousness. So, in many cases communities worked against each other rather than acting together.

In general, decisions in Styria are very centralized, and there is a lack of coordination between local and provincial policy makers. So, no local funds are available, and so local policy makers are not involved in the decision about grants. In contrast, regions and firms have to compete for the funds, which are usually devoted to specific objectives. The provincial authorities think, that these objectives are important for the regional development. This creates several problems. On the one hand, decisions are made by people who do not know the regions well enough, especially in case of extremely peripheral regions like Liezen. On the other hand, firms in more centrally located regions have an advantage. Finally, the system creates an inefficient use of resources, since firms and communities frequently carry out projects not because they are convinced of their importance, but because they can get money for them. In contrast, important projects cannot be realized due to lack of money. This leads to passive behaviour and discourages local initiatives and networks.

(4) Lack of technical education facilities

An important weakness of the region is the lack of technical education facilities. Somebody, who wants to visit a technical school, has to go 1.5 hours by train on average (one direction). High-tech firms will hardly be attracted until this weakness will be removed.

Concluding remarks

Can the lagging region learn from the leading region?

Whether strategies of the leading region could have been similarly successful in the lagging region is hard to determine, since the preconditions were very different. So, Liezen was confronted with a more difficult situation at the outset of the 1980s. Old and large industrial firms in basic industrial branches faced the urgent need to change products and strategies, and to carry through strong rationalization measures. The wage levels were high, and the flexibility of actors was low. In contrast, manufacturing in Osttirol was based on small firms, and wage levels were low. So local actors could concentrate on active employment policies, instead of standing with backs against the wall. Despite those differences, the following can be said:

(1) The province Tyrol has partly decentralized the decision process for the use of regionally important funds, while the province Styria followed a strict method of central decisions. Local funds in Osttirol created high flexibility, and guaranteed the distribution of money according to regional requirements. In contrast, firms in Liezen had to compete for subsidies with firms in other regions, decisions were drawn far from the region, and the room for initiatives of local politicians was much smaller, which might have discouraged the development of active local networks.

(2) In Osttirol local politicians, especially at the level of communities, were more cooperative towards new industrial firms than was the case in Liezen. However, this was not the same in all parts of Liezen. In the east, where firms described local politicians as very cooperative, development was significantly better than in the rest of the region.

(3) In Osttirol the qualification of the labour force was described as one of the main strengths of the region, while firms in Liezen frequently indicated problems in finding qualified people with technical skills. Partly this can be explained by the lack of technical education facilities in Liezen.

(4) In Liezen there is no town that could actually play the role of a regional centre. So, the official regional centre Liezen has neither a hospital, nor a gymnasium, while in Osttirol important institutions like schools, public offices and hospitals are very centralized in Lienz. However, it is hard to say, to what degree this has contributed to attracting firms, or to creating jobs in the public sector.

General lessons

(1) Cooperative behaviour of communities towards companies has been one of the most successful strategies in attracting new firms from outside. However,

even if a lagging region is well advised to follow this strategy, from the overall perspective of rural regions, gains might primarily be at the cost of competing with other rural regions.

(2) Decentralization of funds and decisions seems to be a good strategy in order to improve the competitiveness of rural regions compared with urban regions, since today most decisions are made in urban regions by urban people. It could improve the endogenous potential of the regions, and encourage the development of internal networks. Moreover, it could lead to higher efficiency in public spending.

(3) The investigation of the Austrian case study regions confirms the well-known fact that the development of a good, decentralized medium-skill education infrastructure improves the competitiveness of rural regions. Standardized production will more and more be shifted to low wage countries, and so jobs for untrained workers are supposed to decrease. In contrast, medium level education, especially for technical and business skills, is what firms in rural regions will require in the future, since firms have to be flexible and prove good problem solving capabilities, which is impossible with untrained staff.

References and further reading

Asamer, M., Bratl, H., Fidlschuster, L., Lübke, A. and Reiner, K. (1994) *Regionalwirtschaftliches Konzept für den Bezirk Lienz (Regional development concept for the district of Lienz).* ÖROK-Schriftenreihe, Nr.112, Vienna.

ISIS (Data source of the Austrian Statistical Office ÖSTAT) (1981, 1991) *Volkszählungen (Census).* Vienna.

ISIS (1981, 1991) *Arbeitsstättenzählungen (Firm census).* Vienna.

ISIS (various years) *Fremdenverkehrsstatistiken (Tourism statistics).* Vienna.

ISIS (various years) *Landwirtschaftsstatistik (Agricultural statistics).* Vienna.

ISIS (various years) *Industriestatistik (Statistics of industries),* Vienna.

RISS (Data source of the Austrian Institute of Regional Planning) (various years) *Unemployment statistics.* Vienna.

Tichy, G. (1982) *Regionalstudie Obersteiermark (Regional study for the North of Styria).* ÖIR/WIFO, i.A.d.Bundeskanzleramtes u.d.Steiermärkischen Landesregierung, Vienna.

Weiß, F. (1998a) *Case study Liezen.* Universität für Bodenkultur, Institut für Wirtschaft, Politik und Recht, Vienna.

Weiß, F. (1998b) *Case study Osttirol.* Universität für Bodenkultur, Institut für Wirtschaft, Politik und Recht, Vienna.

Rural Employment in the Arctic Fringe

12

Tuomas Kuhmonen and Olli Aulaskari

Introduction

General overview: harsh climate, scattered settlement, increasing concentration

With an average population density of 17 inhabitants/km^2, Finland is the most rural EU Member State. Due to the tendency of population to concentrate in the agglomerations, towns and village centres of the country, only 19% of the population lives in the really rural areas. In these areas, covering as much as 97.5% of the total area, population density is as low as 3 inhabitants/km^2. Within this settlement pattern, Finnish regions experience varying development patterns. The capital region and the Åland islands between Finland and Sweden generally show the most positive development, followed by the sea coast up to the Oulu town region at the northern end of the Gulf of Bothnia. Most of the regions in these parts have accumulated population during recent decades, whereas the eastern, central and northern parts of the country have been losing their vitality in terms of people and jobs. The flow of people has been from the outskirts of provinces to their capital regions and from all regions to the Helsinki region in the south.

Taking the population base as the most important indicator of vitality of a rural region, the national typology of regions gives a good general idea of the regional development in Finland. Towns and the urban-adjacent rural regions have a positive development, rural heartland regions are gradually losing their vitality, and remote rural regions (incl. islands) are facing difficult and increasing problems (Fig. 12.1). From the general viability point of view, the outlook for the near future of the regions is similar. Population growth has only been forecast for central town regions, but not even for all of them (Fig. 12.2). This is due to the accelerated concentration of people and economic activities at the end of the 1990s, partly caused by the increased international competition as a result of EU membership since 1995, and the related change in many of the domestic policies favouring concentration more than before.

207

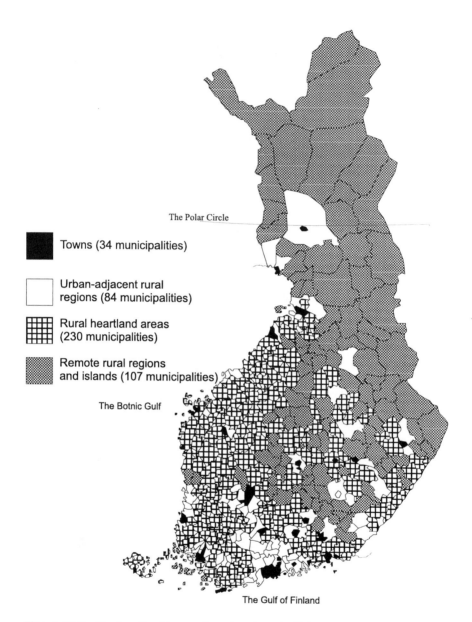

Figure 12.1   Regionalized types of rural regions in Finland
Source: Keränen *et al.*, 1993.

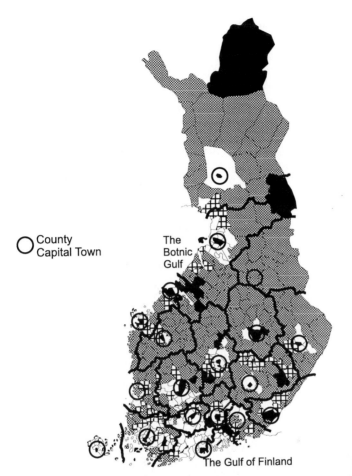

County
Capital Town

The Botnic Gulf

The Gulf of Finland

FORECASTED POPULATION CHANGE IN 1997-2010 (STATISTICS FINLAND)	SHARE OF COMMUTING PEOPLE IN TOTAL NUMBER OF EMPLOYED IN 1995, %	NUMBER OF MUNICIPALITIES
Population will increase	Higher than average (27.7 %)	112
	Lower than average (27.7 %)	34
Population will decrease	Higher than average (27.7 %)	88
	Lower than average (27.7 %)	215

No data (3)

Figure 12.2 Forecasted population change, 1997-2010 and the importance of commuting across municipality borders, 1995

Source: Kuhmonen, 1999.

Scarcity of success stories and the confusing recession of the early 1990s

Out of this diversity of development patterns of different regions, two provinces have been selected for the case studies. The selection is blurred by the deep economic recession in the early 1990s, which caused variation among the regions due to their economic structure (export oriented versus domestic demand based regions). On the basis of employment change in the 1980s and early 1990s, the provinces of Åland (Ahvenanmaa), Uusimaa, Oulu and Keski-Suomi have been successful (above the national average performance). However, Åland is a small semi-autonomous island between Finland and Sweden, and the province of Uusimaa is the province of the capital Helsinki. The success of the province of Oulu is based on the growth of a large town (Oulu) and, in fact, within the province there are large poor rural areas (Kainuu). These special cases are excluded from the selection. So, the only remaining successful rural region is the province of Keski-Suomi, the 'leading' region in the comparison.

Many regions could have been selected as the less successful region for Finland, partly due to the general economic development. The selected lagging region is Mikkelin lääni (province of Mikkeli), a province in the south of Keski-Suomi. The development of non-agricultural employment has been clearly more negative than the national average. The population of the province has decreased by 1.4% in 1980-1995 as well, whereas the leading region simultaneously increased its population by 6.3%. The selected lagging region is neighbouring the leading region of Keski-Suomen lääni in the southeast, and both case study regions are non-coastal, non-border provinces in the central part of the country. They share the same climate, with 150 days of snow cover, average temperature around +3 degrees Celsius and some periods of -20 to -30 degrees Celsius in mid-winter. They are both rather agricultural and have the same kind of regional structure with a relatively large provincial capital, some smaller centres and large rural parts with small municipal centres (as also found in most of the other provinces). This makes it interesting to compare the regions in order to find out reasons for their different development. Both regions belong to the group of most rural regions. They also comprise labour market areas with no significant net commuting across the borders.

Divergence of change in employment

The general change in employment has been similar in most provinces during the last decades: a decline in agricultural and industrial employment and an increase in services (Table 12.1). The performance of industrial employment shows a main difference between the leading and the lagging region. The non-agricultural sectors in the lagging region have not offered sufficient labour demand to compensate for the lost jobs in agriculture and forestry to the extent found in the leading region. The interplay of the agricultural and non-agricultural sector is presented in Fig. 12.3. The small number of regionally limited employment growth regions is evident in Finland. The relatively high share of agricultural employment in the country's extensive rural regions implies a large number of lost jobs in this declining sector and consequently,

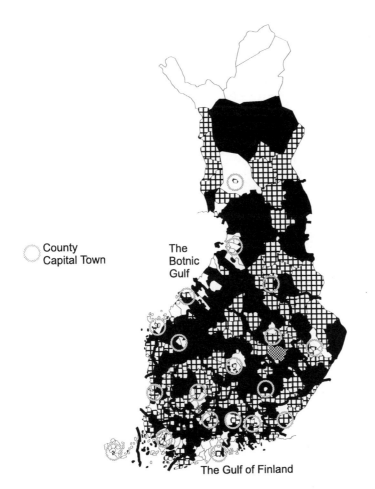

County
Capital Town

The
Botnic
Gulf

The Gulf of Finland

CHANGE IN AGRICULTURAL EMPLOYMENT (INCL. FORESTRY, HUNTING AND FISHING) IN 1975-95	CHANGE IN NON-AGRICULTURAL EMPLOYMENT IN 1975-95	NUMBER OF MUNICIPALITIES
NUMBER OF EMPLOYED DECREASED	Increased enough to compensate the decrease in the agricultural employment	95
	Increased, but not enough to compensate the decrease in the agricultural employment	188
	Number of employed decreased	166
NUMBER OF EMPLOYED DID NOT DECREASE		3

Figure 12.3 Interplay of agricultural and non-agricultural employment, 1975-1995
Source: Derived from Statistics Finland.

Table 12.1   Development of employment and population, 1980-1995 (in % per annum)

	Keski-Suomi	Mikkeli	Finland	Most rural regions
Agriculture	-6.5	-5.0	-4.5	-4.4
Industry	-1.8	-2.9	-2.2	-2.3
Services	0.6	0.6	0.7	0.6
Non-agricultural	-0.3	-0.8	-0.4	-0.5
Total	-1.1	-1.7	-0.8	-1.1
Population	0.4	-0.0	0.4	0.2

Source: Derived from Statistics Finland.

Table 12.2   Economic environment in the Finnish case study regions

	Keski-Suomi	Mikkeli	Finland
Population, 1995 (mill.)	0.26	0.21	5.12
Land area, 1995 (km^2)	16,200	16,300	304,600
Population density, 1995 (inh./km^2)	16	13	17
Population growth, 1980-1995 (%)	6.3	-1.4	6.9
Employment growth 1980-1995 (%)	-16	-22	-11
Population over 64 years, 1995 (%)	15	17	14
GVA growth, 1980-1995 (% p.a.)	6.3	5.4	6.2
GVA per capita (ECU)			
1980	6,400(88)	5,900 (81)	7,300 (100)
1994	12,600(88)	11,400 (79)	14,300 (100)
Structure of employment (%)			
Agriculture and forestry			
1980	20	27	14
1995	8	16	8
Industry			
1980	34	32	35
1995	30	26	28
Services			
1980	47	41	52
1995	62	58	64
Unemployment rate (%)			
1980	5	6	5
1995	21	19	17

Sources: Statistics Finland.

Table 12.3 Population of main cities, 1995

Keski-Suomi		Mikkeli		Finland	
Jyväskylä	74,100	Mikkeli	32,800	Helsinki	525,000
Äänekoski	13,800	Savonlinna	28,900	Espoo	191,200
Jämsä	13,100	Heinola	13,500	Tampere	182,700
Keuruu	12,400	Pieksämäki	13,500	Vantaa	166,500

Source: Statistics Finland.

a great need to create new non-agricultural jobs. However, rural municipalities outside the main town areas failed in this task, with very few (about 10-20) exceptions.

Basically similar types of regions, but structurally still different

Both studied provinces cover approximately 16,000 km^2 and have a population of 0.2-0.25 million people (Table 12.2). The population density is slightly below the national average. Gross value added (GVA) per capita has remained 10-20% below the national average in the 1980s and early 1990s as well. The share of elderly people and the share of primary employment have been higher in the lagging region than in the leading region. Both regions have tended to struggle against relatively high unemployment rates.

One of the most distinct differences between the regions during the period under study is found in their population change. Whereas the lagging region has lost population, the increase in the population base of the leading region is evident. Another distinct feature is the change in employment. In particular, the leading region has managed to maintain industrial jobs much better than the lagging region, despite of their generally declining trend. The regional organization of the population and economic activities is also different. In the leading region about 40% of the population lives in the capital town region around Jyväskylä, whereas in the lagging region this share is only 22%. Consequently, the size of the regional centre in the province of Keski-Suomi is double the size of the regional centre in the province of Mikkeli (Table 12.3). The other few main towns in both regions have a population of 10,000-30,000 and most of the municipalities are very rural with 60-80% of the population living in areas of scattered settlement outside towns or village centres. The general location of both regions in transport logistics within the country is comparable: it takes 3-3.5 hours by car to the Helsinki region, the dominant concentration of the economic activities in Finland with a share of 1% of the land area and 30% of the national GVA in 1993.

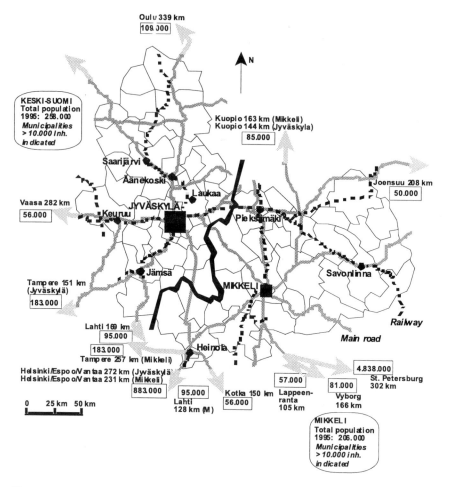

Figure 12.4   Dispersion of population and main traffic lines in Keski-Suomi and
Mikkeli[a)]

a) Distances from Jyväskylä or Mikkeli along main roads. The large towns near the
regions are indicated by the arrows; the number of inhabitants in the box and the
distances from Jyväskylä or Mikkeli along main roads (in km).

Local resources

About 50-140 metres of coastline and 5-6.5 ha of forest land per inhabitant

Figure 12.4 indicates the organization of the population and the main traffic
connections in the case study regions. The stronger concentration of the capital
region in the leading region is evident as compared to the lagging region. Even if the
general location of both regions is comparable in the national scale, the internal

structure is rather different. This is partly caused by the numerous waters. While about 16% of the area in the leading region is covered by thousands of lakes and rivers, the share of waters in the lagging region is as high as 25%, including the largest lake area in Finland (Saimaa). Especially in the lagging region, many areas suffer from distances caused by waters.

Numerous waters are also one of the main strengths of both regions. The amount of coastline in the leading region is 14,000 km, having 27,000 summer cottages in 1994. These are attractive assets, as it is more or less a national habit in Finland to have a summer cottage or a tiny sauna at the lakeside. The respective numbers for the lagging region are 28,000 km and 46,000 cottages. The lagging region in the heart of 'Lake Finland' is the leading cottage and summer habitation region in Finland.

Another important strength of both regions is the abundance of forests, covering over 80% of the land area. These supply raw materials for the timber, pulp and paper industry and generate also 15-20% of the farmers' net income. In 1995, the timber chain from the growing of wood to final products provided 12% of employment in the leading region and 9% in the lagging region. Many of the objective strengths in both regions are related to the agricultural and food sectors, which are declining in employment volumes. The share of the agri-food chain in 1995 was 16% in the leading region and as much as 20% in the lagging region. Despite an extensive summer habitation and a very positive environmental image of the lagging region, its employment performance has been weaker than in the leading region. Besides, the lagging region faced a decline in population. In the next sections we look for possible reasons behind the differential performance in the fields of economic activities, policies and actors.

Economic activities

In the standard nine-sector split, employment increased significantly only in social and personal services in both regions in the period 1980-1994. Besides, there was a small increase in employment in mining in the leading region and in electricity and water supply in the lagging region. In all other sectors employment decreased, partly due to the recession in the early 1990s and partly for structural reasons.

It is possible to disaggregate the change in employment by means of a shift-share analysis, in which the total employment change is separated into a national (general) change, an impact of the sectoral mix and a remaining local share. The national change is the impact of the national average change of total employment in the region concerned (if all sectors and regions shared a same relative change, this factor would count the change as a whole). The sectoral mix counts the impact of the national change related to each sector as compared to the average national change across all sectors. If, for example, a rapidly shrinking sector is dominant in a region, this factor counts the impact. The remaining employment change implies some local impact, either positive or negative. In both regions, the shift-share analysis indicates

Table 12.4 Results of the shift-share analysis in the leading and lagging region: disaggregated change in the number of employed in 1980-94, 9+1 sectors (in thousands)

	State growth		Sectoral mix		Local share		Total change		Employed total, 1980	
	lea-ding	lag-ging	lea-ding	lag-ging	lea-ding	lag-ging	lea-ding	lag-ging	lea-ding	lag-ging
Agriculture, forestry etc.	-2.7	-3.3	-5.2	-6.5	-0.6	-0.6	-8.5	-10.3	17.5	21.8
Mining	-0.0	-0.0	-0.08	-0.1	0.2	0.08	0.06	-0.06	0.2	0.3
Industry	-4.1	-3.0	-4.4	-3.2	0.6	-1.0	-7.8	-7.2	26.8	19.7
Electricity, water supply	-0.2	-0.1	0.06	0.04	0.05	0.09	-0.04	0.02	1.0	0.7
Construction	-1.3	-1.1	-2.1	-1.8	-0.3	-0.4	-3.7	-3.3	8.3	7.0
Trade, accommodation etc.	-2.0	-1.6	0.5	0.4	-0.8	-0.8	-2.3	-1.9	13.1	10.7
Transport, telecomm.	-1.1	-1.1	-0.3	-0.3	-0.8	-0.9	-2.2	-2.3	7.5	7.2
Finance, insurance etc.	-0.6	-0.4	-0.4	-0.3	0.5	0.4	-0.5	-0.3	3.7	2.7
Social and personal serv.	-4.2	-3.3	10.7	8.4	-1.7	-3.3	4.9	1.8	27.4	21.5
Sector unknown	-0.4	-0.4	0.5	0.4	-0.3	-0.2	-0.3	-0.2	2.7	2.4
Total	-16.4	-14.3	-0.7	-2.9	-3.2	-6.6	-20.4	-23.7	108.3	94.0
Total impact shift/share (%)	81	60	4	12	15	28				

the existence of relatively large positive local factors in finance, insurance and estate services and smaller positive local factors in electricity and water supply and in mining (Table 12.4). A major difference between the regions is found in industry, which had a positive local share in the leading region and a large negative local share in the lagging region. In all other sectors both regions shared negative local impacts.

The 'big wheel' dominates local forces

The national general change (recession impact) explains 60% of the apparent employment decline in the lagging region and 81% of that in the leading region. About 12% of the change in the lagging region can be attached to the impact of the sectoral structure of the economy (e.g. share of declining agriculture larger than average), whereas in the leading region this impact was rather small (about 4%). The local share contributed to 28% of the employment decline in the lagging region, but only 15% in the leading region. This share includes the impact of policies, infrastructure, random factors etc., which causes the deviation of the region from the national change, taking into account the impact of the sectoral structure on employment. The shift-share analysis shows that the national development and the regional economic structure are dominating factors with regard to the change in employment in the short and medium run. In the case study regions they explained about 70-85% of the realized change in employment in 1980-1994. So the contribution of the local share, such as the role of local actors or regionally applied policies, is much smaller, but nevertheless of critical importance.

Only small firms increased employment

In the period 1984-1993, employment increased only in small firms with fewer than 10 employees in both regions and declined in larger firms. This was also a national tendency. The enterprise creation rate was clearly higher in the leading region (net change of 2.5% p.a. versus 1.8% p.a. in the lagging region). In both regions the creation rate was highest in towns and lowest in rural municipalities. Most of the 'top ten' employers of the lagging region are part of the public sector, whereas in the leading region the role of the few large private firms is more important (producing e.g. paper machines, tractors, paper and pulp).

Regional settlement pattern matters: central areas versus rural outskirts

In the lagging region, in the period 1980-1995 population increased in eight of the 29 municipalities, but the number of employed only in one of the municipalities. In 1980-95 about two-thirds of population growth in the municipalities of the province (+6,490) took place in the capital town (Mikkeli), 8% in another main town Savonlinna and 8% in Heinolan maalaiskunta (rural district of Heinola), surrounding the main industrial town Heinola. The situation is predicted to remain the same in the forecasts made till 2010, the main accumulation areas being Mikkeli and Heinola (Statistics Finland). Outside the four main town regions with their commuting areas, depopulation occurred (decrease of 10-25% of population) along with a decrease of 30-45% of employed in 1980-95. In the leading region population increased in 12 of the 30 municipalities and the number of employed in two municipalities. In 1980-1995, about one-third of population growth in the municipalities of the province (20,050) took place in the capital town (Jyväskylä), 29% in the surrounding rural commune of Jyväskylä and 13% in the northeastern neighbour municipality Laukaa – all found in the central area of the province.

So in both regions, only towns and their commuting areas have managed to increase their population base. The outskirts of the provinces are faced with depopulation. However, in the leading region, the central sub-region counts roughly half of the population and GVA, whereas in the lagging region the respective figure is 35%. In the leading region the strong and solid concentration of industrial, administrative and other businesses – including a large university – has established the position of the Jyväskylä region as one of the very few national growth poles, whereas such a growth pole is completely lacking in the lagging region. The university of Jyväskylä in the leading region has turned out to be very attractive among students and has thus partly contributed to the very positive national image of the town. Besides, the university affects the supply of labour and know-how for local businesses. The lagging region has only university affiliates.

Agriculture is present in all areas, but dominates especially in the outskirts of the regions. The sector is faced with a difficult problem especially after the EU membership and open competition on the internal market. It is hardly possible to become commercially competitive in basic agricultural production in these areas due to climatic, geographic and structural reasons. On the other hand, it is equally difficult to diversify farm activities and start, e.g. on-farm processing and selling, since a local market for these products does not exist, owing to lack of people or purchasing power in the immediate surrounding in most rural parts. Despite potentials for enlargement, at the moment agrotourism is still limited. Hence, agriculture becomes increasingly dependent on the subsidies to compensate for unfavourable circumstances (lower revenue/higher cost) and agricultural employment will decline rapidly also in the near future.

Actors

External networks have a pronounced role

Actors and their relationships are important in networks. Both internal and external networks affect employment. In both provinces, the external networks are seen as more important for the development of the region than the internal networks. Many preconditions for operation are of a national origin (e.g. labour market regulation, finance of public activities, agricultural policy and forestry policies) and the main concentration of economic activities is situated outside the region (Helsinki area). These have determined the bulk of employment change in both regions. When many firms operate nationally or internationally, the external networks are relatively important in enterprise related connections. These are emphasized especially in the leading region, where approximately 60% of the industrial production is exported (paper machines, tractors, paper and pulp).

The internal networks are well developed, especially in the main town regions and the actors know each other well, especially within the sectoral networks (i.e. industry networks, agricultural networks, forestry networks, policy networks etc.), but between the sectoral networks the situation is not so developed. The internal network between the local university and local firms has been of vital importance in the leading region in order to maintain its competitive position.

In many respects, the engine behind the change in employment in the region comes from internal demographic changes (e.g. internal and external migration, declining agricultural employment) and from external networks setting the frames for economic development (e.g. national economy, public finance) and policies (e.g. agricultural policy). The internal networks are not strong enough, do not have sufficient resources or are administratively not in a position to control these external networks and forces and, therefore, many of their operations are reactions to the activities of the external networks. This setting is more evident in the lagging region than in the leading region, due to more scarce internal economic resources. A lower population density in the lagging region also implies thinner networks and longer distances. However, the differences in the performance of the internal networks do not significantly explain the change in employment in the regions.

Public money flows are important and more beneficial to the leading region

Various policies are of crucial importance for both regions. Redistribution of funds by the State, the National Pension Fund (NPF) etc. has had large regional impacts. This money flow includes a bewildering array of various policy measures like state subsidies to the municipalities to maintain basic services, agricultural policy and traditional regional policy etc. The degree of involvement of the central government is relatively extensive: regionalized state expenditures correspond to 50-60% of the regional GVA in the leading region and to 60-70% in the lagging region. The regionalized state budget indicates higher than national average per capita state expenditures to the leading region, e.g. in the sectors of national defence (material production), education, science and culture (university), social security and health, traffic and labour affairs. The respective expenditures in the lagging region have been higher than average, e.g. in the sectors of agriculture and forestry, traffic and social security and health. The regional economic structure and the facilities of infrastructure also play a role here. The net gain of this public money flow can be defined as state expenditures received by the region less all taxes paid etc. to the State and National Pension Fund. This net gain has been somewhat larger in the leading region, corresponding 11% of the regional GVA in 1994 (9% in the lagging region).

Key factors of success and failure in employment dynamics

When the two case study regions are compared, a number of key factors emerge:

(1) Regional structure and development of the central area

When a significant part of public and private services (having a local and stable demand and slow growth of labour productivity) are located in the central area of the province, the setting of the economic base of this part of the region is also more stable relatively to more industrialized or agricultural areas – especially in a country with a very low population density. Many of the key factors of success in Keski-Suomi are related to the central area around Jyväskylä, the university and regional centre with high R&D expenditures, good airport connections, extensive commuting, more positive enterprise development and the availability of high education. The success in employment of the central area of Keski-Suomi is evident in many sectors as compared to other parts of the province. In more remote parts of the region, the economic structure is more directed to industrial and agricultural activities, with a relatively rapid increase in labour productivity. Without a significant increase in production (which is impossible, e.g. in agriculture), the outflow of labour from these more remote areas is unavoidable. If the annual increase in labour productivity in non-agricultural sectors and the declining trend in agricultural employment are taken into account, the correlation of the regional change in employment and national GDP change suggest that in many parts of the province a very high growth of the national economy would be required to maintain employment (about 8% of GDP volume each year). So the continuous decline of the relatively large primary sector and the effect of a continuous increase in labour productivity produce such a loss of jobs or a need for new jobs as compensation, that the maintenance of employment is very difficult in the more remote areas. In the central region the challenge is much smaller with a GDP growth demand of about 4% p.a., due to a more service-oriented economic structure with a lower increase in labour productivity. So this kind of basic regional structure and sectoral employment pattern have been very significant for the past changes and will continue to be significant for future prospects as well.

In the province of Mikkeli there is no such large dynamic, nationally attractive and expanding centre area/town like in the leading region, which would give better possibilities for networking, synergies and 'critical masses' to develop. There are four main towns in the region and the employment and population development outside the commuting areas of these towns has been extremely negative. The lack of competitive centres has led to significant commuting to two towns outside the region (Lahti and Varkaus). More severe preconditions in the lagging region have resulted in a relatively rapid concentration of population to the main town regions. There are also fewer commuters and fewer extensive part-time jobs, partly due to the regional physical structure and size of the urban centres. The lower population density implies smaller local markets; when compared with Southern Finland or many European regions this is a major difference. Due to the extensive decrease in population, large parts of the region are within decreasing local markets and rather far away from expanding market areas.

(2) A large, attractive local university

The local university in the leading region has attracted students, supported the upkeep of the industrial know-how basis of the region (e.g. in the paper and paper machines production) and contributed to the very positive nationwide image of the central region, especially among young people. The impact is also visible in R&D activities. Despite the presence of the university affiliates in the lagging region, the university based networks and impacts on firms are much smaller.

(3) Favourable industrial structure and competitive industrial performance

The industry of the leading region is relatively export oriented and characterized by the existence of international firms with global markets (Valmet tractors and paper machines, Nokia electronics and mobile phones, forest industries). Many industries are based on its abundant local resources, such as wood, pulp and paper. The key sectors of development are partly related to industries with favourable future perspectives as well (metal, electronics, telecommunications, environment and energy technology etc.). These local factors have positively contributed to employment change. The industrial activity of the lagging region is not very strong. The lower level of education among people becomes a problem when demand for that kind of labour is not extensive in the region.

(4) Participation and commuting possibilities

The large and expanding central area of the leading region has implied growing possibilities for commuting from the neighbouring regions and this has spread the employment impact to a larger area around growth poles. Part-time employment has increased in the region, showing an expansion of more flexible employment patterns, though part-time employment is still not very common. The traffic infrastructure for internal and external movement is relatively good, especially around the central area. The situation of the road infrastructure in the lagging region is more scattered with four smaller main towns and longer distances as a result of water areas (25% of the area!).

(5) Abundant natural resources

Extensive forests have been the basis for industrial activities of both regions in the long term, but the industry in the leading region is larger and much more competitive. Lakes and landscapes are one of the factors attracting existing and new population to stay/settle in the region. The lakes and beautiful landscapes of the lagging region have paved its way to become the leading summer habitation region in the country with a huge number of summer cottages and events, but this has not reversed the other negative developments in employment and population.

(6) 'Great regional policy'

The redistribution of funds by the State, the National Pension Fund etc. has been beneficial to both regions, but more beneficial to the leading region.

Concluding remarks

What can be transferred?

Due to the different circumstances it is not certain that the key factors of success could be transferred to the lagging region in a successful way even with required resources. Besides that, physical settings are hard to change by policy (settlement structure, size of the centres etc.). Some issues can be put forward to improve the employment situation in Mikkeli:

- Certainly, the presence of a local university with many affiliated activities and networks would be beneficial for the lagging region as well (not to forget the direct employment impact with public money, either).
- A competitive industry basis – found in the leading region but lacking in the lagging region – is a result of a long development process with a huge number of contributing factors.
- Maybe further improvement of commuting and participation possibilities also would be possible factors to improve in the medium run, with better traffic infrastructures of all kinds.
- A more extensive use of local, abundant resources is also an avenue ahead, since agricultural products are hardly processed in the region and also forests and lake landscapes could always give basis for higher value added.
- From the non-sectoral point of view, there would be an immediate short-term need to stop the selective outflow (migration) and reduction of population, because otherwise the basis for development diminishes continuously. This is a critical issue in areas where the population density is around 3-5 persons/km^2.

New policy eras?

The existence of regional institutions for internal development and know-how potential seem to be important (e.g. university). Like many other things, this is a policy issue of efficiency/equity to a certain extent as well. Regional centres should be equipped with an institutional structure allowing harmonious development of the key sectors and of the strengths of the region, and allowing achievement of 'critical mass' for becoming competitive and achieving synergies.

The relation between towns and rural regions or rural parts of the regions is important. In Finland some 97.5% of the area comprises scattered settlement and only 2.5% is occupied by cities or villages with dense built-up areas. One-fifth of the population living in strictly defined rural areas should maintain the regional structure, manage natural resources and supply city people with agricultural products and rural services they demand. Due to the ongoing urbanization this relation will gradually change further and in 2020 probably about 10% of the population will live in 97% of the area. Possibilities, amounts and conditions for mutual exchange between rural and urban areas are the key issue for the development of rural areas, as its own demand is very limited and decreasing. Development of this exchange has many forms, such as measures favouring increased commuting (taxes on fuel and cars, deduction of travel costs in taxation, taxation of estates in various types of regions, improved traffic infrastructures and possibly public transport etc.) and measures increasing value added of the exchange in favour of rural regions (increased tourism and leisure activities, processing of agricultural produce and timber etc.).

With regard to lagging regions with relatively limited economic resources and scattered settlement, having suffered for decades from selective out-migration to more competitive regions, the only solution to change the direction of development in the long run would to ask new people to move in. Without reversion of the out-migration, naturally related to the development of jobs, the present indigenous and external policies affecting the region would most probably not have an impact enabling reversion of development. One possibility that has been evaluated as suitable for this could be a special deduction in state progressive income taxation, based on the location of residence in relation to the main national economic centre (national impact), in relation to the main province centre (regional impact) and in relation to the municipal centre (local impact) (Kuhmonen, 1998). Since it is hard to define the activities or sectors of success by external actors, a sufficiently strong incentive could attract and maintain people in this kind of problematic region. The ways to make a living and take advantage of the deduction would be left for them to find out in each case and location. This could be financed by national and/or EU sources. Logically, the same type of direct aid would do the same, but the progressive incentive would then not be possible and the scheme would need much more bureaucracy.

Waiting for the miracle

Anyhow, it seems very difficult to reverse the concentration process on the top of Europe. Many regions are simply falling below the critical line of vitality and only some new direct measures could change their future. Otherwise positive developments happen only in the few large urban growth poles. Because of the specific conditions, the story of the development forces in the Finnish rural regions is to a great extent a story about settlement structures and policies and not so much about actors and networks. It can be questioned if the negative trend of the remote rural regions could be reversed even if Mr Albert Einstein were revived and settled in the middle of the virgin backwoods.

References and further reading

Keränen, R., Malinen, P., Keränen, H. and Heiskanen, T. (1993) *Suomen maaseutulyyppien alueellistaminen (Typology of the rural regions in Finland).* Ylä-Savon Instituutti, Tutkimusraportti 6-1, Sonkajärvi.

Kuhmonen, T. (1997) *Suomen maaseudun elinvoimaisuus ja sen säilyttäminen (The vitality of the Finnish countryside and its preservation).* Fin-Auguuri Oy, Research Reports 2, Vesanto.

Kuhmonen, T. (1998) *Suomen maaseudun kehitys ja kuva vuonna 2020 (The development of the Finnish rural areas and the perspectives for 2020).* Fin-Auguuri Oy, Maaseudun tulevaisuusselvityksiä 3, Vesanto.

Kuhmonen, T. (1999) *Analyysi elinkeinorakenteesta, maatilataloudesta ja maaseutu kaupunki vuorovaikutuksesta Länsi-Suomessa vuosien 2000-2006 aluekehitystyön strategian näkökulmasta (Analysis of the employment structure, farms and rural-urban interaction in Western Finland from the development strategy point of view (Objective 2 programme for 2000-2006)).* Fin-Auguuri Oy, Maaseudun tulevaisuusselvityksiä 5, Vesanto.

Kuhmonen, T. and Aulaskari O. (1999) *Agriculture and employment in the rural regions of the EU - Finnish case studies.* Fin-Auguuri Oy, Future Studies of the Rural Society no 6, Vesanto.

Suomen Kuntaliitto (1997) *Suomi keskittyy ja autioituu; muuttoliike alueittain 1985-1996 (Finland concentrates and becomes deserted; migration in 1985-1996).* Helsinki.

Lessons for Employment Creation in Rural Regions   13

Ida J. Terluin, Jaap H. Post, Heino von Meyer and Béatrice Van Haeperen[1]

Introduction

In the previous chapters a comparative analysis of factors hampering and encouraging the development of employment in nine pairs of leading and lagging regions in different EU Member States during the 1980s and the first half of the 1990s was made. We will now summarize the findings of all 18 case studies by using the 11 key issues, which were formulated in Chapter 2. Derived from this summary and from the comparative analysis, some lessons, which leading and lagging rural regions can learn from each other with regard to employment creation, are given. These lessons are divided into five parts. First, attention is paid to differences in the institutional structure among EU countries, which result in differences in the competences of regions in implementing policies. This elaboration gives rise to recommendations on how to integrate local actors in the decision-making process. Second, a general guideline for employment creation in rural regions is given. Finally, specific lessons with regard to the three components of the field of force (local resources, economic activities and actors) are discussed.

Key issues in the 18 case study regions[2]

(1) Are local resources (including infrastructure) important for the creation of employment?

Almost all case study regions had valuable rural amenities of some sort. Thus, it is difficult to draw any firm conclusion concerning their weight in explaining differential performance in rural employment creation. The comparisons show

---

[1] All members of the RUREMPLO team contributed to this chapter. The authors acted as editors to streamline these contributions.
[2] An extensive report on the comparative analysis of the 18 case study regions is given in Terluin *et al.*, 1999.

that it is not primarily the existence of amenities that matters, but the degree to which these assets are effectively valorized.

Road infrastructure is well developed in all leading case study regions, except for the mountainous parts in some regions, whereas in the most lagging regions road infrastructure is well developed in the central part, but insufficiently developed in the more remote parts. So on the whole in the leading regions road infrastructure is no serious economic constraint for local entrepreneurs. It contributes to an efficient trade of services and goods, and it creates an attractive location for firms to settle. On the other hand, the poor situation of infrastructure hampers the economic development in lagging regions.

(2) In which branches does employment increase/decrease? What are the properties of these branches?

Both in leading and lagging case study regions there was an increase in employment in the sectors of community services and of wholesale and retail trade, restaurants and hotels during the period 1980-1995, along with a decline of agricultural employment. Besides, some leading and lagging case study regions showed also a rise in employment in the financial services sector. The most striking difference between leading and lagging case study regions was the increase in employment in the manufacturing sector in the leading regions, whereas employment in this sector in the lagging regions tends to decline.

The different branches can be classified according to their exposure to global markets, whether the markets are fluctuating or stable and whether they are labour intensive or labour saving. Employment growth in leading regions is not dependent on certain properties: it increases both in branches exposed and less exposed to global markets, in fluctuating and stable markets and in labour-intensive and labour-saving branches. However, in lagging regions employment mainly increases in branches characterized by less exposure to global markets, stable markets and labour intensive production. So in leading regions employment development is relatively more vulnerable than in lagging regions.

(3) Does the sectoral mix explain the dynamics in employment growth/stagnation?

Do leading regions have an underrepresentation of employment in shrinking sectors like agriculture and industries and an overrepresentation in expanding sectors like services? If sectoral employment structures do not differ among regions, the residual explaining divergence in employment development can be labelled as 'territorial dynamics'. This is supposed to reflect specific regional characteristics. In most of the leading and lagging case study regions the sectoral shares of agriculture and industries in employment exceed those of the national economy. Based on such a sectoral mix, a below average growth should be expected. This indeed occurred in the lagging regions. However, it did not happen in the leading regions, which implies that territorial dynamics are an explaining factor of employment growth rather than the sectoral mix.

(4) Is employment created in small or large enterprises?

Both in leading and lagging case study regions employment growth takes place in small enterprises. In some regions employment growth in medium and large enterprises is reported. However, growth in medium and large enterprises occurs more often in leading regions than in lagging regions.

(5) Is employment created in new or existing enterprises?

Whether employment is created in new or existing enterprises seems to be affected by country-specific factors, rather than by being a leading or a lagging region. So for most countries it was found that employment in leading and lagging regions is created mainly in new companies, while for a few countries the analysis showed that employment was created in both existing and new firms.

(6) Does the education level of the labour force matter in the creation/ stagnation of employment?

On the whole it can be stated that in leading case study regions the education level of the labour force is relatively low. However, the employment structure is such, that this type of labour is demanded and the abundance of low-skilled labour is a pull factor for industrial firms to settle in these regions. In the lagging case study regions it was often reported that both poorly and well educated labour was available. So from the labour supply side education was no constraint on employment growth in lagging regions. In some lagging regions the lack of employment opportunities for highly educated people resulted in out-migration.

(7) Is employment hampered by the institutional structure of the labour market?

A first impression is that the institutional setting of the labour market in leading and lagging case study regions does not differ from other regions in the country, since it is determined at the national level. Hence, minimum wage levels apply for the whole country, and are no specific constraint or incentive for employment growth in the case study regions. The picture of the role of employment services/agencies in matching supply and demand of labour varies: in some regions they show a good performance and in others they are insufficient. However, the performance of employment services/agencies is not related to the status of being a leading or lagging region. In some regions matching of supply and demand often takes place in an informal way, which reduces the role of employment services/agencies.

(8) Does the capacity of actors matter in the creation/stagnation of employment?

Capacity can generally be defined as the ability of actors to cooperate and interact in the market and usually refers to the three aspects of knowledge, skills and attitude. In most of the leading case study regions the capacity of policy makers is rather well developed, whereas in most of the lagging case study regions the capacity of policy makers is rather weak. Positive aspects in the capacity of policy makers in leading regions are the way in which they implement policies according to the priorities and needs of the region, in which they are able to attract public funds and private investments and in which they create preconditions for firm settlement. Weak points in the capacity of policy makers in lagging regions refer to inability to formulate strategies, lack of political consensus, lack of good contacts with upper level authorities and inability to identify the needs and priorities of the region.

In a number of leading and lagging case study regions the capacity of entrepreneurs is well developed. This is often the result of a restructuring process in traditional industries. The new and small companies are competitive in national and international markets. However, their capacity to innovate is often limited. In other leading and lagging case study regions the capacity of entrepreneurs is weak, due to a cautious and risk-averse attitude or to lack of industrial tradition.

The capacity of labourers seems to be roughly the same in leading and in lagging case study regions: their attitude to work is good and they are prepared to work hard.

(9) Specify the role of internal and external networks in the creation/ stagnation of employment

On the whole leading case study regions were characterized by rather strong internal networks, whereas those in the lagging case study regions were usually rather weak. The internal networks in the leading regions were for example enhanced by an active attitude of local actors, solidarity, easy communication and strong local leaders. Problems faced in the internal networks in the lagging regions are a low density of actors, little interaction among internal actors, a lack of cooperation among sectors, internal conflicts, lack of active actors, lack of capacity of local actors and lack of formal networks, which are able to guide the development process.

External networks are considered here to be the interactions of actors inside and actors outside the region. It appears that the most frequent use of external networks is to get financial support from regional/national/EU level (policy relations), to export products (market relations) and to be in contact with (multinational) firms, either due to the presence of subsidiary business in the region or to attract firms into the region (firm relations). In the leading case study regions external networks functioned better than in the lagging case study regions. Difficulties in the external networks of lagging case study regions are due to the marginal/remote position of the region within a larger administrative unit, lack of unified strategies, lack of capacities of the local actors and an inward-looking attitude of the local actors.

The engine of employment growth consists of a mix of endogenous and exogenous forces in all case study regions, except for Pesaro and Macerata. In these regions, which belong to the so-called 'third Italy', industrial districts exist and endogenous forces are the engine of employment growth. It is striking that in leading regions endogenous forces tend to initiate the process of employment growth, which is subsequently enhanced by exogenous forces. In lagging regions it was often found that exogenous forces tend to initiate the process of employment growth, and that endogenous forces react on them.

(10) Give an identification of the most effective policies and strategies towards maintaining or augmenting employment

In both leading and lagging regions strategies of policy makers were directed towards the improvement of infrastructure, financial support to firms, setting up of public services, improving the education level of the labour force and supporting economic activities in thinly populated areas. A main difference in the strategies of policy makers in leading and lagging case study regions was that policy makers in leading regions were more often involved in setting up industrial sites with appropriate equipment, compared with policy makers in lagging regions. The advantage of such industrial sites is that these can create synergy effects. In some lagging regions strategies of policy makers were weak due to the failure to include these in a broader development perspective.

Although companies are a direct source of employment, usually the purpose of a firm is not to create employment but to make profits. A common strategy for firms in both leading and lagging case study is to improve their competitiveness in the market by higher quality products, technological innovation and flexibility. In some leading and lagging case study regions a tendency to self-employment can be perceived.

(11) How do farm households adapt to the situation of decreasing employment in the agricultural sector?

One of the results of the decline of the agricultural labour force is that land becomes available for farmers, who continue farming. So in all case study regions, except for the Austrian ones, the main adaptation strategy of farm households is farm enlargement in the sense of increasing the land area per farm. In some leading and lagging regions this strategy was combined with an intensification of production, due to the use of new techniques such as irrigation or large-scale machinery. Another main element in adaptation strategies is the shift from bulk production to niches (products of regional origin), high quality products and organic farming.

The level of pluriactivity is dependent on the availability of jobs in the regional economy, the demand for products processed at farms, the demand for services like agrotourism and nature conservation provided by farmers and country specific factors. The three most common forms of on-farm pluriactivity are agrotourism, processing and selling of farm products and forestry. Off-farm pluriactivity refers to a great variety of jobs in the industries and services sector. It is remarkable that in the case study regions in Greece, Italy and Spain farm households are rarely involved in on-farm pluriactivity.

Due to the presence of landscapes of outstanding scenic beauty or high natural value and other rural amenities in the case study regions, farm tourism offers promising perspectives as a source of income. In leading case study regions farm tourism is more common than in lagging case study regions, Osttirol and Liezen being the exceptions. Problems faced in developing farm tourism in lagging regions are the lack of a regional strategy towards tourism and farmers' lack of knowledge about agrotourist opportunities. In some regions like Drenthe and Liezen a saturation level has been reached and hence prospects for agrotourism are in particular in a shift towards high quality accommodation. The current participation in agri-environmental programmes in countries like Austria and Germany is quite high, while it is virtually non-existent in Greece, Spain and Italy. The future uptake of these programmes depends mainly on the size of the premiums.

Institutional structures and employment dynamics[3]

Institutional patterns, and in particular administrative structures and procedures, are important aspects in any assessment of regional development dynamics. The responsibilities and tasks of the various administrative levels and bodies involved in rural or regional policy design and implementation vary significantly among EU Member States. In some countries, regions have a great deal of autonomy, in others they are just administrative bodies implementing national policies. The analysis below shows that the embeddedness of local actors in institutional structures and procedures can be improved. Table 13.1 provides a rough overview on the regional administrative structures for the nine RUREMPLO countries. While some countries like France and Spain show a regionally deep-structured administration, others, like the Netherlands or Finland have only one administrative layer between national and municipality level. The size of the regions selected for the case studies ranges from 50,000 (Osttirol) to about 1 million inhabitants (Niederbayern), the majority, however, being in the range of 200,000-300,000 inhabitants. In some case study regions there is a regional government or council, which is directly elected by the local population (indicated by bold letters). However, in case study regions of Germany, France, Italy, Austria and Finland, the regional authorities are appointed by authorities of the superior (Germany, France, Italy and Austria) or of the inferior (Finland) sub-national level.

Table 13.2 provides an impression of the domains that fall within the competences of the case study regions. Case study regions at a lower territorial level, like the Districts in Austria, have few matters to decide, while others, like the Regierungsbezirke in Germany, have major responsibilities in shaping regional/rural policies. In most cases, the administrative layer corresponding to the territorial entity of the case study region is competent in providing basic infrastructure and is strongly involved in land-use planning. With respect to regional development funds, only a few case study regions are involved in decisions about projects (Netherlands, Finland and Germany). Other case study

---

[3] This section is derived from Von Meyer *et al.*, 1999.

Table 13.1 Sub-national administrative structure in the countries of the case study regions

Population in thousands (indicative) [a]	Belgium	Germany	Greece	Spain	France	Italy	Netherlands	Austria	Finland
>10000	State	State	State	State	State	State	State		
5000-10000		Region (Land)						State	State
1000-5000	Region			Region	Region	Region			
500-1000		Regie-rungs-bezirk	Region					Pro-vince	
100-500	Pro-vince		Prefec-ture	Pro-vince	Depart-ment	Pro-vince	Pro-vince		Pro-vince
20-100	District	Kreis			Canton			District	
<20	Munici-pality	Munici-pality	Munici-pality	Munici-pality	Munici-pality	Munici-pality	Munici-pality	Munici-pality	Munici-pality

a) The shaded ranges of the administrative layers refer to our case studies. We have to read this table in this way: the population of the French departments selected for the case studies ranges between 100 and 500 thousands inhabitants; these departments belong to regions where population ranges between 1 and 5 million inhabitants.

regions are only involved in administrative tasks. With respect to horizontal funds, the case study regions are left with very few or no competence at all, except for Germany. The same is true for regulations concerning labour contracts, job mediation and unemployment benefits etc. that in most countries appear to be more centralized.

Diversity of administrative structures and employment development

The diversity in administrative structures among countries deserves the following comments:

(1) The pairs of case studies – one lagging and one leading region for each country (except for Belgium) – face of course the same countrywide institutional structure. The institutional structure of the country can therefore not explain the differences in the development of employment in the lagging and leading regions. What matters in the explanation is not the design of the institutions – centralized or decentralized – but the relations between local actors and these institutions. The lack or low level of official competence of some case study regions regarding some issues – for example the use of regional EU funds – does not imply that local actors are excluded from the decisions. In some case study regions, for example Luxembourg (B), formal and informal networks of local and external actors succeed to design and prepare specific projects, which they present to the upper level authorities responsible for final decision. Therefore, in cases where competence is above the regional level, networks of local actors with upper administrative layers

Table 13.2 Competence of the administrations in the case study regions

	Belgium	Germany	Greece	Spain	France	Italy	Netherlands	Austria	Finland
Level	Province	Regierungsbezirk	Prefecture	Province	Department	Province	Province	District	Province
Regional funds (EU)	•	••	•	•	•	•	••	•	••
Regional funds (national/regional)	•	••	•	•	•	•	••		••
Horizontal funds (EU)		••	•	•	•	•			•
Horizontal funds (national)		••	•	•	•	•	•		•
Road infrastructure	••	••	••	•	••	•••	•	••	••
Social and health infrastructure	••	•••	••	••	••	••			••
Education	••		••	••	••	••			••
Spatial planning	••	••	••	•	••	•••	•••	••	••
Labour contracts			•	••					
Job mediation		•	•	•			•	•	••
Unemployment benefits		•	•	•	•		••	•	•
Other labour market policies		••	•	••	••	•	••	••	••
Agricultural policies		••	•	•	••			•	•

The points are used in the following way. One point indicates that the case study regions are only involved in pure administrative tasks, two points, that they are also involved in decisions about projects and funds and three points demonstrate even an involvement in the design of the rules.

must be facilitated so that local actors can affect/participate in the decision-making process. The other way to involve local actors in the decision process is to apply the subsidiarity principle in such a way that local actors are included.

(2) The diversity of the administrative structure among countries has implications as concerns the delivery of EU policies. These policies have to be designed in such a way that all rural regions are able to fully exploit the potentiality of the measures offered. At this regard, information plays a major role. It is essential to ensure wide promotion and extension of EU regional

measures. All actors actually or potentially concerned with the development of rural regions must be correctly and fully informed about all measures. While in some dynamic rural regions, local actors are integrated in well informed internal and external networks, in other rural regions, local actors are not. The following recommendations should help in achieving these goals:

- to implement in the EU regions one 'information point (one stop shop)' providing information and help about the EU regional measures;
- to simplify the procedures and reduce the paper burden;
- to decide the contents of EU policy in mutual consultation with the regions;
- to involve groups of competent local actors in the supervision of EU regional projects.

General guideline for employment creation

Based on the comparative analysis of the case study regions, we now formulate some lessons for stimulating employment in rural regions, again by using the three main components of the field of force: local resources, economic activities and actors. As the three components are strongly interrelated, lessons often concern aspects of the other components as well. Since the socio-economic, physical and geographic situation of rural regions varies widely, there is no one unique development path towards more jobs. So the lessons formulated below should not be considered to be the 'success formula', which always results in more jobs. The lessons have to be seen as building blocks, which may contribute to shaping preconditions for employment creation under certain circumstances.

Despite the multiple development trajectories, we give a general guideline for employment creation in rural regions, based on the experience in the case study regions:

(1) make a comprehensive territorial development plan, based on the strengths, weaknesses, opportunities and threats of the region, and integrate all measures and projects within the scope of this plan;

(2) improve the capacity (knowledge, skills and attitude) of local actors;

(3) strengthen the cooperation of local actors and the cooperation of actors inside and outside the region.

Regional administrative layers and entrepreneurs are the main actors in implementing the three elements of the guidelines. In many cases encouragement from upper administrative levels will be required. Within the framework of this guideline the following lessons – if suiting the needs of the region – may be selected and applied. In the lessons no attention is paid to the way in which these lessons have to be implemented, since that is beyond the aim of the project.

Lessons with regard to local resources

Integrate infrastructure investment in a broader development process

Physical infrastructure is an important factor for rural development. The case studies show that investment in infrastructure alone is not sufficient to trigger positive rural development. It will not in itself create employment opportunities, except during the (short) construction period. Comparison of the case studies provides evidence, that in the longer run infrastructure investment management makes a significant difference. In several case study regions improved connections to major transportation networks inside and outside the region have been essential for making transport of products and services more efficient. In most regions efforts have also been made to create new industrial sites, equipped with water treatment plants and other infrastructure facilities. This suggests that infrastructure investments should be integrated into a broader comprehensive development concept, and be accompanied by a set of complementary incentives. Such a comprehensive development concept should be based on a systematic assessment of regional strengths and weaknesses, as well as future opportunities and threats.

Pay attention to distinct types of infrastructure in rural regions

In improving infrastructure and providing public services, it must also be recognized, that in order to be efficient, rural regions often require types of infrastructure and technologies distinct from those in urban regions. Explicit consideration of rural characteristics and needs is demanded, e.g. in providing public transport, health care, education, or sewage treatment.

Valorize rural amenities

Almost all case study regions have some valuable rural amenities, which contribute to their 'local identity'. However, the existence of these amenities are not able to explain employment dynamics, but the degree to which these assets are managed and valorized by actors to generate added value and employment. Rural amenities have to be managed in such a way, that the sustainability is not endangered.

Improve the perception of amenities by rural actors

There is often a gap in the perception of rural amenities by rural people and that by people outside rural regions. An important precondition for valorizing rural amenities is that rural actors are conscious of the values of rural amenities, i.e. that they understand that unspoiled nature, attractive landscapes, historic villages, etc. are scarce resources and unique development assets, that should be kept in good shape. This is not only a service for tourists and leisure-seeking urban populations. The consciousness of living in a unique village may have spin off effects for the rural population as well, as it can break a negative circle and result in new energy and activities. Rural renewal schemes can help to initiate such processes.

Lessons with regard to economic activities

Follow a multisectoral approach

Rural employment creation results from complex processes of economic growth and decline, structural change, adjustment and innovation. The case study regions showed an increase in employment in the branches of community services, wholesale and retail trade, restaurants and hotels and financial services during the period 1980-1995, along with a decline of agricultural employment. Besides, several case study regions showed also a rise in employment in the manufacturing and construction sectors. Policies aiming at encouraging rural employment creation should follow a multisectoral approach, mainly by shaping preconditions for local agents.

Support the integration of agriculture in the rural economy

In the more thinly populated parts of rural regions the decline of the agricultural labour force may endanger its viability. In order to maintain agricultural workers in those areas, additional employment opportunities have to be created outside the farm or employment opportunities on the farm have to be stimulated, such as the production of public goods (nature conservation) or agrotourism. A main obstacle for developing agrotourism is the lack of a regional tourist strategy.

Both specialization and diversification can be successful strategies

The leading case study regions provide evidence that both specialization and diversification can be successful strategies. There are, however, no typical rural specializations, which could be predefined a priori. Some of the leading case study regions are typical examples of so called 'industrial districts' (e.g. Pesaro, Albacete, Luxembourg (B) which, due to an exceptional specialization of their economic system, even manage to compete on an international, global scale. So despite the above lesson on a multisectoral approach, a certain degree of specialization can be useful. Of course, regional specialization is not without risk. Market conditions, tastes and fashions change. Under such circumstances rural employment policies should help to anticipate change and adapt to new conditions. It cannot be said, however, if further specialization or diversification are generally the right choice. This is underlined by the fact that other leading rural regions have been successful by diversifying their economic base. Often they show above average growth across all major branches. Although diversified regions appear to be less exposed to risks, they may, however, find themselves more exposed to competition from other rural or urban regions if they lack proper market niches, a clear regional profile and image that can be easily communicated.

Enhance facilities for new and small enterprises

The case studies show that a substantial part of employment is created in new and in small enterprises. This implies that policies should not focus only on existing and large enterprises, but rather enhance facilities for new and small enterprises.

Focus on the local productive system

The case studies show that success and failure do usually not depend on the location and investment decisions of individual firms. What matters is the functioning of the entire local productive system, which results from the interaction of a multitude of firms as well as other institutions and actors. So rural employment policies should avoid targeting individual firms exclusively. Economic development and employment growth can benefit when chambers of commerce, local banks or other institutions manage to offer managerial training, transfer of technological and organizational know-how, advise on investment and financing, in a way that is adapted to the needs of small rural enterprises.

Strengthen zoning of economic activities by spatial planning

It appears that firms and actors tend to move to towns and agglomerated parts, which reflects the attractiveness of concentrations of actors. Such concentrations often result in synergy effects. Spatial planning can be used as a policy instrument to enhance this concentration of activities by providing well equipped business sites in certain zones. Natural locations for such concentrations of activities are towns, waterways or motorways. In a number of regions larger towns are lacking, which often hampers economic development. In order to create a structure with some larger towns, spatial planing can be used by focusing on the creation of business sites in one or two villages/towns of the region. A concentration of economic activities in some parts also provides the advantage that it contributes to the safeguarding of the attractiveness of rural amenities and living conditions in other parts of the region.

Lessons with regard to actors

Enhance capacity building of local actors

In our field of force we distinguish three components: local resources, economic activities and actors. The overall finding is that actors are the essential and decisive factor in rural development. The key question with regard to the actors is whether they have the capacity (knowledge, skills and attitude) to take the right steps towards encouraging employment. This capacity depends on the degree in which actors face their situation and prospects in the broader national and international context. So policy makers have a high capacity when they have the ability to act effectively in delivering policies, to support promising local initiatives and projects and to formulate policies to attract investments. Entrepreneurs have a high capacity when they have the ability to perceive changes and adjust to them, and when they show the willingness to respond to market changes. Labourers have a high capacity when they have the ability to adapt to changes and to adjust their skills to training needs.

From the case studies it appeared that in most of the leading case study regions the capacity of policy makers is rather well developed, whereas in most of the lagging case study regions the capacity of policy makers is rather weak. Key issues in the capacity of policy makers are:

- political consensus;
- the ability to make a diagnosis of the regional situation, to identify needs and priorities, and to plan and design appropriate projects within a comprehensive territorial development perspective;
- the way in which policy makers are able to have good contacts with upper level authorities;
- the way in which policy makers are able to have good contacts with entrepreneurs;
- the way in which policy makers are able to attract public funds and private investments;
- the way in which policy makers create preconditions for firm settlement.

Entrepreneurs operate in the market. When they do not have the capacity to adapt to changing market conditions, they will not survive in the long run, unless they are supported by public assistance. In a number of both leading and lagging case study regions, weak points in the capacity of entrepreneurs were a limited capacity to innovate and a cautious and risk averting attitude.

The capacity of labourers seems to be roughly the same in leading and in lagging case study regions: their attitude to work is good and they are prepared to work hard. In situations of restructuring of traditional industries groups of fired labourers, who lack the capacity to adjust their skills in order to be employed in higher skilled jobs, became permanently unemployed. So training of labourers is a main target point.

Strengthen internal and external networks

A network is considered here to be a group of actors, who interact with each other in order to achieve some aim. On the whole leading case study regions were characterized by rather strong internal and external networks of policy makers and entrepreneurs, whereas those in the lagging case study regions were usually rather weak. Target points for actions towards strengthening networks are enhancing the solidarity and interaction among local actors, improving the cooperation among sectors, solving of internal conflicts, stimulating an active attitude of actors, preventing an inward looking attitude of the local actors and encouraging the interaction of internal and external actors. It is clear that these target points are closely interrelated with empowerment of the capacity of actors.

Attract newcomers

The case studies show that newcomers to rural regions, immigrant populations, entrepreneurs and policy makers from outside the region, or even tourists can play an important role in establishing external links. Local actors, who have stayed outside the region for a long time, and return to the region, can also be counted in the group of newcomers. Due to the fact that newcoming people have a different attitude from the local actors, they are able to mobilize the local actors. They can feed experiences into internal networks, help mobilize local actors and act as local leaders. They can provide access to external know-how and markets. They can transport a positive regional image, which supports advertising and marketing of local products.

Define the right labour market area

Rural employment policies depend on a proper functioning of regional labour markets. Rural employment policies have to facilitate the matching of regional labour supply and demand. In order to design targeted regional measures it is essential to have a clear understanding of what represents the actual labour market area. This is not self-evident because of changing commuting patterns, increasing travel to work distances and changes. Besides, the labour market area may differ for different professions. Often, administrative boundaries no longer reflect actual functional relationships. Thus, a precondition for any targeted rural labour market policy is to get a clear picture of what represents the relevant labour market area. This implies also an understanding of the role of regional centres, small and medium size towns in providing job opportunities for populations living in the countryside. Exchange of vacancies between employment services of neighbouring regions can facilitate the matching and supply of labour.

Aim at the appropriate regional mix of skills

Education and training play of course an important role in matching labour supply to demand and thereby in encouraging employment creation. The role of education is however highly complex. It is not the attainment level as such, but rather an appropriate regional mix of skills that matters for successful rural employment growth. Proper targeting of education and training is required to ensure a better regional balance. For example, in those regions where employment growth was particularly high in manufacturing industries, greatest demand was expressed for workers with medium level technical skills. Establishing technical schools and promoting professional training both within and outside enterprises are priorities. Employers themselves were interested in providing professional qualifications to manual workers by on-the-job training. In those regions where strong regional networks and partnerships existed, the matching of skills seemed to work particularly well.

Be aware of changes in labour demand by industrial firms

During the period we have analysed, the availability of unskilled labour often acted as a pull factor for industrial firms to settle in rural regions. However, due to competition with cheap firms in low wage countries, it can be wondered whether footloose industrial firms will stay in rural regions in the near future. It is probable that industrial firms in rural regions will change their production in such a way that they become more flexible, service intensive and customer oriented. Such a shift could imply that industrial firms will prefer medium skilled labourers to unskilled workers.

Encourage part-time labour and self-employment

In many rural regions young and female populations are particularly affected by unemployment. Regional labour market policies should thus pay particular attention to their specific needs. Where employment opportunities are limited, it seems urgent to think not only in terms of full-time hired employment, but also to consider alternative options such as job-sharing, flexible part-time arrangements or self-employment. Part-time labour, pluriactivity and self-employment have a long tradition in many rural labour markets. The majority of farm families in Europe are used to such work and income patterns. For the development of many rural labour markets these traditions can be a positive advantage. It is probably not by accident that some of the most dynamic rural labour markets, showing the greatest relative employment increases are those with high shares of pluriactive, part-time farms. In many industrial districts the work ethic and attitudes of workers, who have strong ties to traditional pluriactive farming systems, are important for explaining their success. Many rural regions have a long tradition of independent self-employment. This should be encouraged again. Risk taking is not new to many rural people. As a result, new forms of organizing economic activities can actually find positive preconditions in rural regions.

References

Terluin, I.J., Post, J.H. and Sjöström, Å. (eds) (1999) *Comparative analysis of employment dynamics in leading and lagging rural regions of the EU, 1980-1997.* LEI-DLO, The Hague.

Von Meyer, H., Terluin, I.J., Post, J.H. and Van Haeperen, B. (eds) (1999) *Rural employment dynamics in the EU; Key findings for policy consideration emerging from the RUREMPLO project.* LEI-DLO, The Hague.

Index